The Fight Against Climate Change: Exploring Carbon Capture Solutions

Youshva

Contents

5 Summary and Outlook

Appendix

Chapter 1

Introduction

1.1 Motivation

After the worldwide financial crisis in 2009, the annual global primary energy consumption has consecutively climbed for ten years to approximately 580 exajoules in 2019 [1]. Fossil energy, such as oil, coal, and natural gas, still holds dominant shares in the primary energy supply, with respective 33.1%, 27.0%, and 24.2% in the global energy mix [1]. The utilization of fossil energy inevitably leads to greenhouse gas (GHG) emissions, especially carbon dioxide (CO_2) being the major contributor. Scientific evidence revealed an apparent correlation between the atmospheric greenhouse gas concentration and the global temperature rise (known as global warming) [2]. Increasing primary energy consumption and long-term global warming result in substantial climate changes, which are concerned with human Well-Being and sustainable development of society and economics. Under this background, on 12th December 2015, at the twenty-first session of the Conference of the Parties (COP 21), a universal agreement to limit the global temperature rise was adopted by 195 countries. The Paris Agreement's central aim is to strengthen the global effort to address climate change issues by keeping a global temperature rise in this century well below $2\,°C$ above pre-industrial levels and pursuing to limit the temperature increase even further to $1.5\,°C$.

To reduce GHG emissions, carbon capture and storage (CCS) technologies have been developed with retrofitting of the existing fossil-fuel-fired power plants. Multiple technical solutions were proposed for CCS, such as pre-combustion capture, post-combustion capture, oxy-fuel combustion, and chemical looping, to name only a few [3]. Previous studies have been substantially reviewed in the literature [3–7], exploring concepts, challenges, and urgent demands for research, spanning from fundamental studies to industrial applications. Among the several proposed technical solutions for reducing CO_2 emission, oxy-fuel combustion is a promising technology for coal-fired power plants [5]. This approach suggests using pure molecular oxygen (O_2) as the oxidizer produced by air separation and eliminates nitrogen (N_2) in the combustion process. Consequently, the flue gas mainly contains CO_2 and water (H_2O), facilitating CO_2 capture and storage. Flue gas recirculation regulates the flame temperature within a range comparable to conventional air conditions and avoids combustion boilers' damage.

The oxy-fuel combustion is reported as the most energy and cost-efficient approach among the CCS technologies [5]. However, the combustion processes are significantly impacted by introducing CO_2 as the primary inert species due to its chemical and thermal properties. For example, the molar heat capacity, diffusivity, emissivity, and density of CO_2 are mostly different from N_2. These properties impact the combustion behavior, such as ignition, flame propagation, and flame stabilization in CO_2 environments. The solid fuel combustion itself is a multi-phase and multi-parameter physico-chemical process, adding further complexity for the practical implementation of oxy-fuel combustion. A better understanding of oxy-fuel combustion requires a detailed study of the underlying sub-processes, which should be individually analyzed under well-controlled experimental conditions, e.g., in generic systems. With a step-wise increase in complexity, single particle combustion (SPC) and particle group combustion (PGC) should be addressed with experimental investigations. Combustion measurements on the particle level provide deep and clear insight into sub-processes and their behavior in CO_2 atmospheres, allowing for the understanding of real systems. Particle-flame interaction, particle-particle interaction, and particle-flow interaction are essential aspects of laboratory-scale experiments. A better understanding of solid fuel combustion bridges the gap between research and application and strengthens fundamental knowledge to reduce or even eliminate CO_2 emissions of pulverized-coal-fired power plants.

1.2 Aim and Structure of the Work

1.2.1 Aim

Solid fuel combustion includes numerous complex physio-chemical sub-processes. Experimental investigations are needed to understand the individual sub-process and their interactions to optimize the oxy-fuel combustion in pulverized-fuel-fired (pf-fired) power plants. Experimental studies should start with laboratory-scale combustion systems, and vast fundamental knowledge obtained from there could improve industry-scale applications. With the focus on laboratory-scale experiments, this work aims to characterize high-volatile bituminous (hvb) coal particles' combustion behavior in oxygen-rich environments and acquire a deeper understanding of oxy-fuel combustion fundamentals.

For this purpose, this section identifies several requirements guiding the experimental work. Concerning multiple essential quantities (i.e., scalars and velocities) involved in solid fuel combustion study, an in-situ acquisition of multiple parameters is not replaceable. The acquisition methods should be ideally non-intrusive to avoid adding bias into the experimental data. Considering the time and length scales of the transient processes in coal boilers, experiments need to provide sufficient time and spatial resolutions. It request developing and implementing high-speed volumetric diagnostics. Additionally, experiments should be able to provide data to validate models and compare them with the simulations. Therefore, the accuracy of applied experimental methodology and the freedom for a systematic parameter variation are essential aspects. The following sections elaborate these considerations from different perspectives to make this present work's global aim clear.

Diagnostics Development

In studies of the multi-phase transient solid fuel combustion processes, non-intrusive laser diagnostics for simultaneous scalars and vector measurements are desired. Diagnostics such as particle image velocimetry (PIV) and laser-induced fluorescence (LIF) have been widely used in two-dimensional scalar and vector measurements. However, turbulent flames involved in most practical combustion boilers are inherently three-dimensional and evolve in time. Due to the unresolved out-of-plane information, the 2D measurements might lead to data interpretation ambiguities. This work makes particular efforts to extend existing laser diagnostics for performing 3D and 4D measurements. This advancement of volumetric imaging techniques should demonstrate its capability and feasibility in studying well-known reactive flows. After thorough validations against established methodologies, multi-parameter and multi-dimensional optical diagnostics with high accuracy and precision must be applied to study solid fuel combustion.

Process Focus

In the present work, one of the scientific perspectives emphasizes the ignition and volatile combustion characteristics, which are essential for the stabilization mechanism of coal-fired flames. For the bituminous coal and biomass particles, the ignition and volatile combustion are dominant gas-phase processes due to their high volatile content; hence, experiments are mainly scoped on the gas-phase phenomenon. Besides, investigations should prioritize the single particle combustion at high heating rates and also consider group particle effects, which reinforce basic understanding. Physical parameters of the solid phase, e.g., particle size, shape, and particle number density, need be addressed as well. To understand the interactions between gas and solid phase, atmospheres with defined thermochemical states are preferred, and the interference of turbulent mixing should be rather eliminated. Therefore, the experiments are conducted in a laminar flow reactor supported by a burner stabilized premixed CH_4 flat flame, which is capable of reproducing oxy-fuel-relevant atmospheres. The feasibility of this configuration has been examined in previous studies [8–10].

Boundary Conditions and Parameter Variations

Experimental studies should not solely be used for phenomenological investigations but also for model validation and to develop or evaluate more complex numerical simulation tools. In addition to data acquisition accuracy, well-defined boundary conditions are crucial for a justified validation with numerical simulations. A reasonable comparison between simulation and experiments can only be drawn with identical boundary conditions. Inlet parameters, such as gas temperature, gas velocity, gas composition, and particle velocity, should be thoroughly characterized in experiments. A substantial parameter variation in the combustion atmosphere, e.g., gas composition and oxygen excess, should be taken into account to reflect different regimes in a realistic pf-fired combustion chamber. Moreover, particle size, particle number density, and fuel type are considered, and their influences on the ignition, volatile combustion and flame stabilization demand an in-depth analysis.

Based on scientific perspectives mentioned above, the superior attention of the present work is twofold: (1) advancement of multi-parameter and multi-dimensional laser diagnostics for possible 4D measurements of pf-fired turbulent flames, and (2) investigations of ignition and volatile combustion of solid fuel particles by applying the developed methodology in a generic combustion environment with defined boundary conditions and substantial parameter variations. According to these two perspectives, this work's structure is outlined and introduced in the next section

1.2.2 Structure

The present study is structured in the framework of a cumulative **book**. An introduction of the theoretical background is given in Chapter 2, in which fundamentals regarding laser diagnostics and solid fuel combustion are explained. Then, Chapter 3 and 4 summarize methods and selected results based on six peer-reviewed journal publications included in this work. These publications can be divided into two major parts, as illustratively shown in Fig. 1.1. While in part one, new developments of novel volumetric laser diagnostics are discussed, their applications in solid fuel combustion are emphasized in part two. Brief excerpts highlighting the experimental methodology and scientific outcomes are present for each publication, as shown in Fig. 1.1. The number of each paper is indicated on the right-top corner of each study corresponding to the paper list. To address the scientific perspectives discussed in Section 1.2.1, these attempts start from basic ideas and gradually proceed with increasing complexity. This principle of "from simple to complex" is applied throughout the investigations; thus that obtained findings, experiences, and insights could be further utilized for the next steps.

In part one, studies focus on laser diagnostics' advancement to perform multi-parameter and multi-dimensional optical measurements for combustion research. The methodological development starts with a single-shot 3D reaction zone visualization using volumetric laser excitation (**paper I**), extends to a quasi-4D flame topology measurement using a rapid laser scanning technique (**paper II**), and finalizes with a quasi-simultaneous high-speed 3D scalar and vector measurement campaign by combining both multiple sheets and volumetric illumination approach (**paper III**). These volumetric diagnostics demonstrate the experimental capabilities for analyzing the flame topology, stabilization mechanisms, and flame-flow interactions in various premixed, non-premixed, and partially premixed flame configurations. Part two emphasizes on the applications of the above-mentioned optical diagnostics to solid fuel particle combustion and understanding the ignition and volatile combustion process in oxy-fuel conditions. Herein, the complexity step-wise increases from 2D measurements of single burning particles (**paper IV**) to 3D measurements of single particle (**paper V**) and particle group combustion (**paper VI**). In the following, individual experimental studies and their correlations are summarized in brief.

In Chapter 3, following a short literature review in Section 3.1, the published work of **paper I** [11], **II** [12], and **III** [13] are summarized in three application examples. Section 3.2.1 (refer to **paper I**) evaluates a new approach for 3D flame structure diagnostics using the tomographic laser-induced fluorescence of the OH radicals (OH-LIF). It combines the vol-

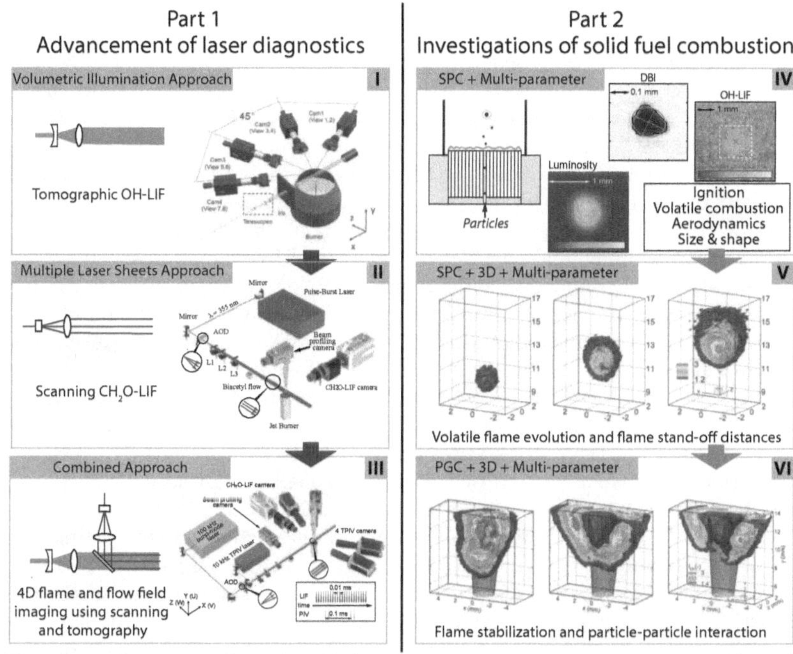

Figure 1.1: Primary results and highlights of the publications included in this work [11–16].

umetric laser excitation with a multi-camera detection system of eight views. Single-shot tomographic OH-LIF measurements were performed in a methane/air premixed laminar flame and a non-premixed turbulent CH_4 jet flame. The number of views and their angular orientation, and the size of the volumetric illumination are evaluated concerning the tomographic reconstructions' accuracy. Using an alternative method for 3D imaging, Section 3.2.2 (refer to **paper II**) introduces high-speed volumetric CH_2O-LIF measurements using a pulse-burst laser operated at a repetition rate of 100 kHz. Quasi-4D CH_2O-LIF imaging at a scan frequency of 10 kHz is realized with a novel laser scanning system employing an acousto-optic deflector (AOD). The diagnostic capability of time-resolved volumetric imaging is demonstrated in a partially premixed DME/air lifted turbulent jet flame near the flame base. Based on the findings in both studies, the laser scanning method is selected for 3D scalar imaging, and the volumetric laser illumination is chosen to perform 3D velocity measurements. Section 3.2.3 (refer to **paper III**) demonstrates high-speed scalar-velocity field measurements in a lifted partially-premixed dimethyl-ether/air jet flame using simultaneous CH_2O-LIF and tomographic PIV (Tomo-PIV). The 3D LIF measurements are performed by scanning the laser beam from a 100 kHz pulse-burst laser rapidly across the probe volume using an acousto-optic deflector. The volumetric reconstruction of LIF signals from ten parallel planes provides quasi-instantaneous 3D flame structure visualization. The temporally resolved flame surfaces and velocity field data are used to analyze Lagrangian particle trajectories and displacement speeds at the lifted flame's base.

Chapter 4 presents first a short review of previous optical experiments on solid fuel combustion with relevant configurations. Then, laser diagnostics discussed above are implemented in three applications, namely, **paper IV** [14], **V** [15], and **VI** [16]. In Section 4.2.1 (refer to **paper IV**), a multi-parameter study of single coal particle combustion in laminar flow conditions is introduced, including a high-speed OH-LIF, luminescence imaging (LU), and diffuse backlight-illumination (DBI) measurement. Simultaneously acquired experimental data allow for evaluating particle size, ignition delay time, and volatile combustion duration for individual particles in the air and oxy-fuel atmospheres. Based on a close comparison with numerical simulations, particle temperatures, local gas temperatures, and fuel mass fraction are evaluated, providing insights into the devolatilization and volatile combustion fundamentals. Extending to 3D, Section 4.2.2 (refer to **paper V**) describes an experimental investigation using high-speed scanning OH-LIF to study igniting particles by temporally tracking OH-LIF signals of the gas-phase flame. The three-dimensional OH-LIF signals were used to reconstruct the volatile flame structure of burning particles, while the particle size and location are measured by DBI. Single particle combustion study is highlighted by evaluating the spherical volatile flame's temporal evolution and the flame stand-off distance. Section 4.2.3 (refer to **paper VI**) presents a volumetric OH-LIF measurement using an AOD scanner combined with a time-resolved DBI providing fundamental insights into the flame topology. Compared to Section 4.2.2, three-dimensional visualizations of volatile flames mainly characterize the transition from single particle to particle group combustion. Combining an in situ determination of the particle number density, the critical particle loading density or inter-particle distance for a significant particle-particle interaction is evaluated.

Conclusively, the results are summarized, and the outlook for future work is presented in Chapter 5.

Chapter 2

Theoretical Background

This chapter aims to present a concise introduction to the theoretical background which helps understand the physical phenomena discussed in Chapters 3 and 4. This introduction starts with some fundamental aspects in solid fuel combustion in Section 2.1, followed by a description of relevant laser-based optical diagnostics in Section 2.2.

2.1 Solid Fuel Combustion

2.1.1 Solid Fuels

Fossil solid fuels can be characterized depending on the degree of carbonization or coalification. Coalification refers to a natural evolution process in which buried plant matter evolves into a dense, dry, carbon-rich, and hard material. Coalification is a function of heat and pressure acting over time, whereas heat is considered as the primary factor [17]. Depending on the degree of coalification, four major ranks of coals are defined, namely, lignite (often referred to as brown coal), sub-bituminous coal, bituminous coal, and anthracite, whereas peat is usually considered to be the precursor to coal. The different ranks vary in color and hardness, but most fundamentally in elemental composition. The evolution in composition from biomass through the peat to different coal stages is illustrated in the Van Krevelen diagram shown in Fig. 2.1 [18]. In this figure, atomic H/C and O/C ratios reveal very high values for the solid fuel types with a low degree of carbonization, i.e., biomass and peat. In turn, both ratios evidently decrease with an increase in the coal rank.

Apart from the elemental composiion (often shown in the ultimate analysis), differences in solid fuels are evidenced in the proximate analysis, in which the fixed carbon, volatile material, ash, and moisture content are characterized. As a consequence, the heating values change with the coal ranks. Figure 2.2 [20] illustrates the correlation between the volatile mass ratio and the low heating value (LHV) for various types of solid fuels. Both anthracite and bituminous coals have higher LHVs than other fuels, whereas lignite reveals lower LHVs with comparatively higher volatile fractions. Biomass contains the largest fraction of volatile material of generally above 60%. Among various fuel types mentioned above, bituminous coal, lignite, and torrefied biomass are experimentally investigated in the present work.

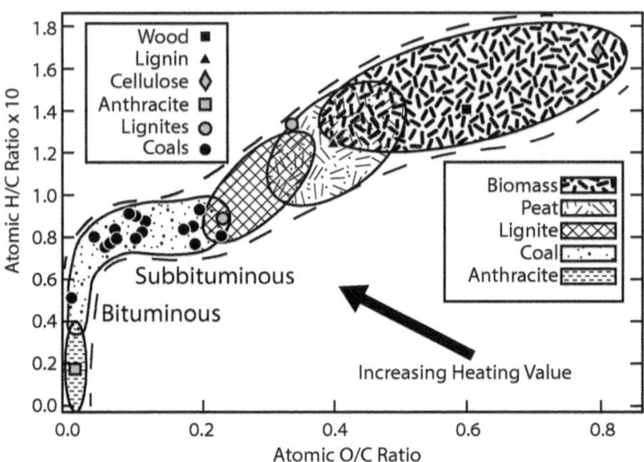

Figure 2.1: Van Krevelen diagram for various type of solid fuels, adapted from [18, 19].

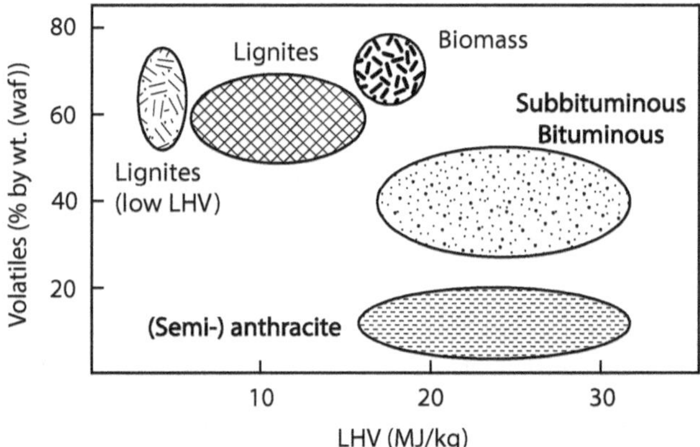

Figure 2.2: Schematic correlation between volatile mass fractions and low heating values for various solid fuels, adapted from [20].

2.1.2 Single Particle Combustion

Consider an individual solid fuel particle with a diameter d_p and a temperature T_p of 300 K being suddenly exposed to a high-temperature oxidizing environment. The particle undergoes rapid heating due to the convective and radiative heat transfer followed by subsequential devolatilization and combustion stages. Depending on the particle size, flow conditions, and fuel types, the duration of each phase and the accompanying particle temperature rise could be substantially different. A homogeneous gas temperature $T_{\mathrm{g},\infty}$ of approximately 1800 K and constant gas velocities are assumed, which are also ensured within the laminar flow reactor investigations in this work. Under these preconditions,

Figure 2.3 schematically presents the temporal behavior of the particle and gas temperature rise. Regarding the high-volatile fuels involved in this work, several essential stages are described in the following.

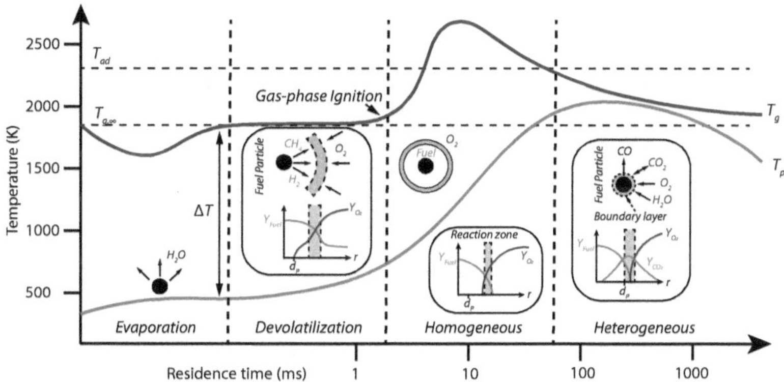

Figure 2.3: Schematic of the particle temperature and gas temperature history in different stages of single particle combustion, adapted from [21].

Particle Heating

Particle heating continues throughout the entire combustion process. The heating rate is a crucial quantity to understand pulverized coal combustion at the single-particle level. However, it is strongly influenced by particle size, thermal convection, and radiation near the solid-gas interface. In the pre-ignition stage (includes evaporation and devolatilization), the current gas temperature T_g is much higher than T_p. The heat transfer from the gas to the solid phase leads first to moisture evaporation, which slightly cools down the surrounding hot gases. Thus, T_g slightly decreases from $T_{g,\infty}$ defined by the given boundary condition. Subsequently, devolatilization happens upon increasing T_p, and different fuel gases are released into the gas phase. The particle heating rate essentially affects the duration of evaporation, the volatile release rate and the composition of released fuel gases. Considering the balance of the internal energy and heat transfer, temporal variations of the particle temperature can be described by the following expression assuming a spherical particle (adapted from [22]):

$$\frac{\pi}{6}d_p^3\rho_p c_p \frac{\mathrm{d}T_p}{\mathrm{d}t} = \pi d_p^2 h(T_g - T_p) - \pi d_p^2 \sigma(\varepsilon_g T_g^4 - \varepsilon_p T_p^4) + Q_{\mathrm{vap}} + Q_{\mathrm{vol}}, \qquad (2.1)$$

where d_p, T_p, ρ_p, c_p and ε_p stand in sequence for the diameter, temperature, density, heat capacity and emissivity of particles; h is the convection coefficient of heat transfer; T_g and ε_g denote the temperature and emissivity of gases, respectively, and σ is the Stefan-Boltzmann constant. The Q_{vap} and Q_{vol} are respectively due to evaporation and volatile reaction, and the latter is often considered negligible, especially prior to ignition.

The convection heat transfer coefficient h scales linearly with the Nusselt number Nu, a ratio of convective to conductive heat transfer from the gas to the solid phase. It is characterized by Ranz and Marshall [23] for the forced convection of spheres:

$$Nu = 2 + 0.6Re_{\mathrm{p}}^{\frac{1}{2}}Pr^{\frac{1}{3}}, \tag{2.2}$$

where Re_{p} is the Reynolds number based on the relative velocity (also referred to as slip velocity) and Pr is the Prandtl number. Pr characterizes the ratio of momentum diffusivity to the thermal diffusivity:

$$Pr = \frac{c_p\mu}{\kappa}, \tag{2.3}$$

in which μ and κ stand for dynamic viscosity and thermal conductivity, respectively. The Re_{p} is defined as:

$$Re_{\mathrm{p}} = \frac{(u_{\mathrm{g}} - u_{\mathrm{p}})d_{\mathrm{p}}}{\nu_{\mathrm{g}}}, \tag{2.4}$$

where u_{g} and u_{p} denote the gas and particle velocity, respectively, and ν_{g} is the kinematic viscosity of gases.

Devolatilization

Devolatilization of coal particles refers to as a process in which coal is heated to yield volatile matter and produce solid residues. Coal devolatilization is often termed pyrolysis when it happens in an inert gas environment. Generally, both expressions are not strictly distinguished from each other due to similar volatile composition and char chemistry. The products from the coal devolatilization contain gases, tar, and char. The gas formation usually relates to the thermal decomposition of large aromatic molecules. However, tars, often defined as room-temperature condensibles released from the solid components at elevated temperatures, undergo more complex physical and chemical transformations [24]. The volatile yield greatly depends on the heating conditions and the coal ranks. A general devolatilization model including several sub-steps was introduced by Solomon et al. [24] based on a bituminous coal particle, as schematically summarized in Fig. 2.4.

In the first stage of metaplast, the devolatilization starts by rupture of the weakest bridges in the molecular structure, especially the C-C bridges between aromatic rings. This process, often referred to as depolymerization, results in forming different small fragments making up the metaplast and free radicals producing fuel molecules in the gas phase. Residue solid compounds present a basic for char formation. Then, due to the decomposition of functional groups in the primary pyrolysis, different gases such as CH_4, CO_2, H_2, and light aliphatic gases are released. Besides, small fragments are released from the metaplast as tar by evaporation and diffusion, while large fragments in the metaplast undergo recondensation or re-polymerization and retrograde to char. Tar has been seen as the precursor for the soot particle formation. It should be noted that coals with different

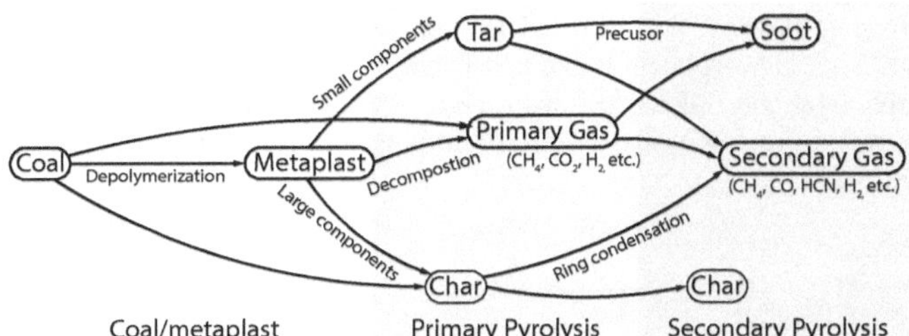

Figure 2.4: Schematic of pyrolysis steps for a bituminous coal particle (adapted from [24, 25]).

molecular structures behave differently during the primary step. In the secondary pyrolysis, additional CH_4, CO, HCN, and H_2 are released due to cross-linking ring condensation [24]. During devolatilization, fuel-rich gas mixtures are formed in the near field of the particle and fuel-lean in the far-field. Figure 2.3 also implies that fuel gases and oxidizers diffuse into each other to form a flammable mixture around the particle.

Gas-phase ignition

For the bituminous coal particle with high volatile content, two combustion stages can be characterized (see Fig. 2.3), namely, the homogeneous and heterogeneous combustion. The homogeneous combustion, also referred to as gas-phase combustion, occurs upon igniting a volatile and oxidizer mixture; in opposition, the following heterogeneous combustion of char is dominated by surface reactions at elevated temperatures. During devolatilization, a flammable gas mixture is formed by the molecular diffusion of volatile and oxidizer into each other. The flow motion could facilitate mixing. In a laminar condition, an annular area with flammable mixture fractions can be expected in the vicinity of the particle, as illustrated in Fig. 2.3. Auto-ignition occurs at a proper temperature of the gas mixture. For auto-ignition following the classical thermal explosion problem, the ignition delay time τ_i of a reactant gas mixture can be expressed as [26]:

$$\tau_i = \frac{c_v T_0^2 / T_a}{q_c Y_{F,0} A \exp(-T_a/T_0)}, \tag{2.5}$$

where c_v denotes the specific heat capacity of the mixture; q_c is the reaction heat release rate; T_0 is the initial temperature, T_a is the activation temperature; $Y_{F,0}$ is the fuel mass fraction; and A is the pre-exponential factor for reaction kinetics. The ignition time increases with the specific heat, and has a sensitive dependence on T_a and T_0. The T_a is defined as

$$T_a = \frac{E_a}{R} \tag{2.6}$$

with E_a being the reaction activation energy.

Following the gas-phase ignition, the volatile matter is oxidized in a thin layer, assuming finite-rate reactions. The volatile flame is a typical diffusion flame, in which the reaction rate is controlled by the mixing/diffusing rate as well as the devolatilization rate. The volatile reaction rate can be written in Arrhenius form with a simplification to a one-step global reaction [6]:

$$\frac{\mathrm{d}c_{\mathrm{F}}}{\mathrm{d}t} = c_{\mathrm{F}}A\exp(-T_{\mathrm{a}}/T_0),\tag{2.7}$$

where c_{F} denotes the molecular concentration of fuel gases.

2.1.3 Particle Group Combustion

Particle group combustion is a more complex multi-phase process in which the heating, devolatilization, ignition, and combustion of individual particles strongly interact and behave differently to an isolated particle. Figure. 2.5(a) shows a schematic illustration of particle group ignition and combustion proposed by Annamalai et al. [27, 28]. Herein, for simplification, a spherical cloud of a radius R_{c} is considered. A group number G is defined to characterize the number density of particle clouds:

$$G = 2\pi n R_{\mathrm{c}}^2 d_{\mathrm{p}},\tag{2.8}$$

where n stands for the number of particles per unit volume, and d_{p} is the diameter of mono-sized particles. The region inside the cloud radius R_{c} is a two-phase zone, while the region outside is a one-phase zone. After introducing a high ambient temperature, the thermal conduction, radiation, and convection (if present) result in increased gas and particle temperature within the particle cloud. Particles near the cloud boundary are first heated by absorbing thermal energy. Hence, heat penetration towards the cloud core are gradually depressed in the radial direction. Obviously, the mean radial gas temperature prior to ignition depends on the number density or the inter-particle distance L_x. Upon the devolatilization temperature, particles release volatiles diffusing along the radial direction into the one-phase zone. A flammable gas mixture can be expected at a certain radius outside of the particle cloud by mixing with the oxidizer. Following a spontaneous ignition at elevated temperatures, a volatile-fueled enveloping flame is formed. Analogous to the isolated particle, a diffusion flame is sustained with fuel-rich mixtures in the near-field of the particle cloud and fuel-lean mixtures in the far-field (see Fig.2.5(b)). The flame could stabilize or move towards the particle depending on the balance of fuel and oxidizer diffusion rates. The gas temperature increases after the ignition supplying more enthalpy to particles for devolatilization. After the consumption of the volatile matter, char combustion takes place with an increasing surface temperature.

This fundamental consideration of a spherical particle cloud neglects the slip velocity and non-uniform particle sizes; hence, it is not directly applicable to study the coal-fired combustors involving dense clouds in turbulent flows. Nevertheless, it can be used for a basic comparison with the single particle combustion, which is often assumed to have

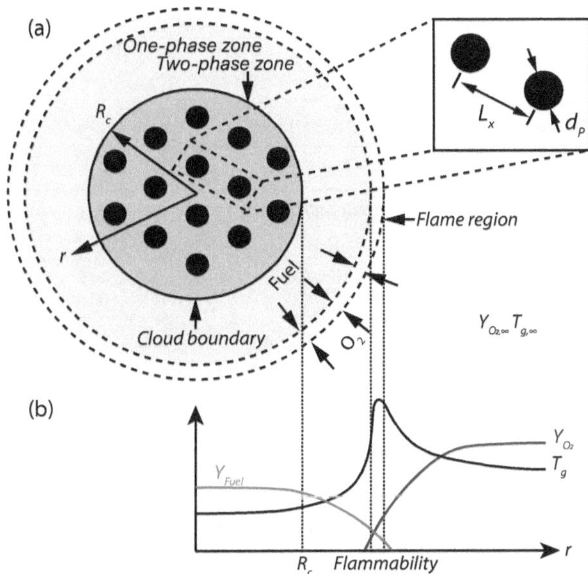

Figure 2.5: Schematic of particle group combustion, adapted from [27, 28].

a spherical shape in simulation and modeling. A basic understanding of the heat and mass transfer and the gas-phase reactions can be achieved based on this model, which is valuable for a proper assessment of complex systems.

2.1.4 Oxy-fuel Combustion

In oxy-fuel combustion, solid fuels' oxidization occurs in pure oxygen instead of air producing CO_2-rich flue gases. After the removal of water vapor, the flue gas is readily prepared for CO_2 sequestration and storage. Hydrocarbon combustion in pure O_2 results in very high flame temperatures, which could introduce irreversible damages to the boiler. In coal-fired power plants, recycled flue gases are suggested for the boiler inlet flow to reduce the flame temperature. The flue gas containing CO_2 and H_2O introduces numerous changes in the combustion behavior due to different physio-chemical properties as opposed to N_2. Thus, oxy-fuel combustion shows distinct characteristics compared to traditional air-firing combustion. Changes mostly relate to two major properties: the radiative property and the heat capacity [3].

Unlike N_2, triatomic molecules such as CO_2 and H_2O have high thermal absorptivity and emissivity. CO_2 and H_2O concentration in oxy-fuel combustion are higher than that in traditional conditions, increasing the radiative heat flux even if the temperature is unchanged. Additionally, the thermal radiation from soot and char particles can influence the radiative heat transfer. The soot formation reveals a sensitive dependence on the local mixing of fuel and oxidizer in coal flames. Due to the changes in gas composition and flow dynamics in oxy-fuel combustion, soot formation and its thermal radiation behave differently.

Table 2.1: Properties of gases at 1400 K and atmospheric pressure [5].

	H_2O	O_2	N_2	CO_2	CO_2/N_2
Density (ρ) [kg/m^3]	0.157	0.278	0.244	0.383	1.6
Thermal conductivity (k) [W K/m]	0.157	0.278	0.244	0.383	1.2
Specific heat capacity (c_p) [kJ K/kmol]	45.67	36.08	34.18	57.83	1.7
Specific heat capacity (c_p) [kJ K/kg]	2.53	1	1.22	1.31	1.1
Heat sink (ρc_p) [kJ K/m^3]	0.397	0.278	0.298	0.502	1.7
Dynamic viscosity (μ) [kg m s]	$5.02{\times}10^{-5}$	$5.81{\times}10^{-5}$	$4.88{\times}10^{-5}$	$5.02{\times}10^{-5}$	1.0
Kinematic viscosity (ρc_p) [m^2/s]	$3.20{\times}10^{-4}$	$2.09{\times}10^{-4}$	$2.00{\times}10^{-4}$	$1.31{\times}10^{-4}$	0.7
Mass diffusivity of O_2 in X (ρc_p) [m^2/s]	-	-	$1.7{\times}10^{-4}$	$1.3{\times}10^{-4}$	0.8

Table 2.1 lists several key thermodynamic properties of H_2O, O_2, N_2, and CO_2 at 1400 K and atmospheric pressure [5]. CO_2 has a much higher heat capacity compared to N_2, which affects the gas temperature and further the gas-phase ignition and flame stabilization. For instance, the adiabatic flame temperature (AFT) sufficiently changes if keeping the ratio between oxygen and "inert" gas constant. Wall et al.[4] calculated the theoretical O_2 volume fraction required to reach the same AFT in oxy-fuel and air conditions, shown in Fig. 2.6. It was concluded that the O_2 mole fraction is about 28% for the wet cycle and 35% for the dry cycle to achieve the same AFT. Assuming the same O_2 mole fraction, a temperature drop of about 400K and 800K is expected for the wet and dry cycle, respectively.

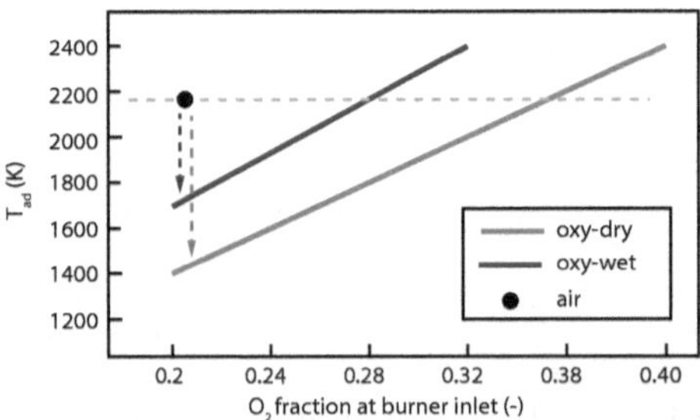

Figure 2.6: The O_2 fraction required at burner inlet to achieve similar adiabatic flame temperature as the air-fired case for wet and dry flue gas recycle, adapted from [4].

Besides, convective heat transfer alters in oxy-fuel combustion due to a different temperature gradient and the coefficient h; the latter is a function of flow dynamics and gas properties such as viscosity, thermal conductivity, heat capacity, and density, which collectively change with the N_2 replacement. The O_2 diffusivity in CO_2 is also lower than in N_2, which affects the mixing and the flammability.

2.2 Laser-based Optical Measurements

2.2.1 Laser-induced Fluorescence

Theory

Laser-induced fluorescence is a resonant process of light-matter interaction, in which electrons are first excited to a higher energy level by absorbing photons and then relax back to lower energy levels with radiative emission termed fluorescence. Figure 2.7 illustrates the LIF process using an adapted Jablonski diagram, where the ground state and the excited state are denoted as S_0 and S_1, respectively. For the OH-LIF and CH_2O-LIF processes involved in the present work, the S_1 stand for the first excited electronic state. In reality, the molecule's energy structure is often complex due to the vibration and rotation motion of two or more atoms. In addition to electronic states, which reveal relatively large energy gaps in the order of several electron volts (eV), there are vibrational and rotational energy states with smaller energy gaps between each other. The energy splitting between electronic states is one order of magnitude greater than the splitting of vibrational levels and three orders of magnitude greater than rotational splittings. The population of different energy levels is a function of temperature and determined by the Boltzmann distribution.

A molecule in the ground state can absorb an incident photon, whose energy matches a discrete energy gap between two energy states of electrons, and is instantly excited to a higher energy level. This excitation transition is often realized by a narrow-band laser emitting photons with the desired amount of energy $h\nu_i$, h and ν_i being the Planck's constant and optical frequency, respectively. Due to the vibrational and rotational energy transfer (VET and RET), the excited electron spontaneously relaxes to lower energy levels of the excited state; and this process is also termed the internal conversion (IC). With a finite time (in the order of 10^{-13}s) after the excitation, the electron relaxes back to the ground state, emitting a radiative fluorescence photon with a frequency ν_f. Due to the energy loss, the emitted photon's wavelength shifts towards larger numbers, which is referred to as Stoke-shift or red-shift. The whole process completes within a short period in the order of $10^{-8} - 10^{-9}$ s, and the fluorescence has a short lifetime in the order of nanoseconds.

Additional to the transition from S_1 to S_0 emitting fluorescence photons, electrons can relax from S_1 to another excited triplet state T_1 with lower energy. This process is often referred as to inter-system crossing (ISC). Electrons relax from T_1 back to the ground state and emit phosphorescence photons. Compared to fluorescence, the laser-induced phosphorescence is a long-time process ($\sim 10^{-3}$ s depending on the particular molecule) and has a lower emission intensity.

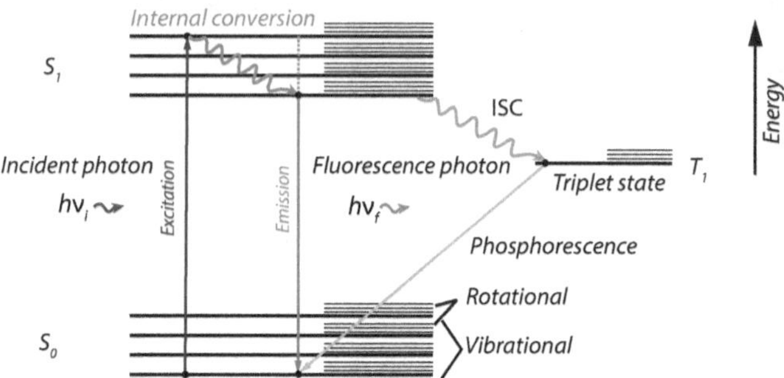

Figure 2.7: Adapted Jablonski Diagram.

Quantification of OH-LIF signals

There are several competing processes for the OH-LIF such as stimulated emission, collisional quenching (Q), photoionization, and predissociation (P) etc., and some of them are shown in Fig. 2.8 by using a simple two-level model. The possibility of spontaneous fluorescence emission by energy relaxation from the excited state to the ground state is described by the quantum efficiency ϕ, which is defined as

$$\phi = \frac{A_{21}}{A_{21} + Q_{21} + P} \tag{2.9}$$

where B_{12} is absorption rate, B_{21} is stimulated emission rate, A_{21} is the Einstein coefficient, Q_{21} is quenching rate, and P is predissociation rate.

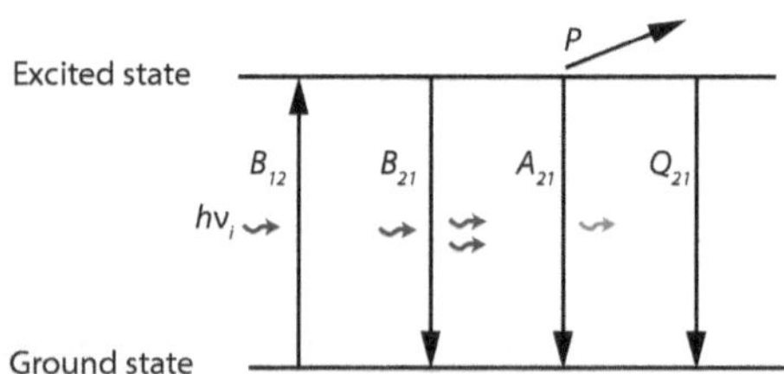

Figure 2.8: Two-level model for laser-induced fluorescence and its competing processes.

Application of OH-LIF

Quantifying the OH mole fraction requires the corrections of interfering processes, making quantitative concentration measurements very challenging [29]. However, the LIF has proven to be a powerful technique to measure species' existence due to high sensitivity and selectivity. For example, the OH-LIF is a well-established approach for flame front visualization [30] and temperature measurements [31] in combustion studies. OH is a highly reactive radical which forms in the high-temperature region of the most hydrocarbon-fueled flames. The electronic transition $(A^2\Sigma^+ \leftarrow X^2\Pi)$ is commonly used for OH-LIF measurements. With respect to the $A^2\Sigma^+(\nu' = 1) \leftarrow X^2\Pi(\nu'' = 0)$ transition used in the current work, Figure 2.9 presents detailed rotational structures within two vibrational levels, namely $\nu'' = 0$ in the ground state $(X^2\Pi)$ and $\nu' = 1$ in the first excited electronic state $(A^2\Sigma^+)$ [32]. Between these two vibrational states, the energy gap is about 35.3×10^3 cm^{-1}, which accordingly demands an excitation wavelength around 283 nm. The excitation wavelength is usually achieved by the frequency doubling of the selected stimulated emission at 566 nm of Rhodamine 6G excited by a Nd:YAG laser at 532 nm. Due to VET and RET, excited OH molecules will also populate lower vibrational-rotational levels, e.g., in $\nu' = 0$. Thus, fluorescence is emitted from several rotational levels in $\nu' = 1$ and $\nu' = 0$ of the $A^2\Sigma^+$ state and the wavelength spans over a narrow range centered at 309 nm.

Figure 2.9 additionally indicates a commonly used excitation line denoted as $Q_1(6)$. The transitions between two rotational levels create different transition branches. There are two energy ladders in the $\nu'' = 0$ level. According to the difference in quantum number J, the branches are termed O, P, Q, R, S, with ΔJ being -2, -1, 0, 1, 2 in the order. For the $Q_1(6)$ transition line, the number in brackets describes the number N of the ground level, and the subscript indicates in which energy ladder the transition occurs. More details about the molecular energy structure refer to [33, 34].

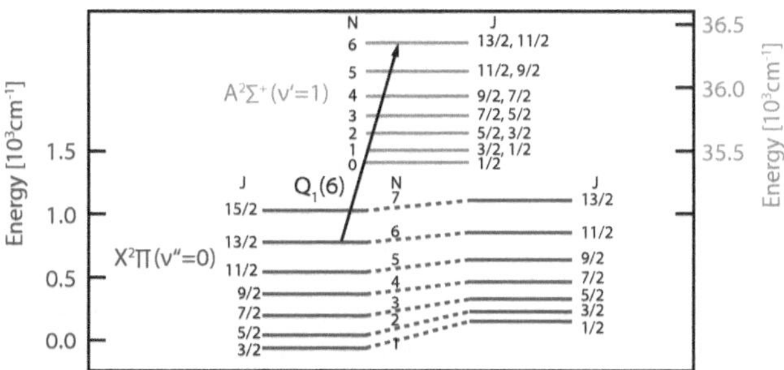

Figure 2.9: Retational energy levels in two vibrational state $\nu'' = 0$ of $X^2\Pi$ and $\nu' = 1$ of $A^2\Sigma^+$. $Q_1(6)$ excitation line is highlighted, adapted from [32].

2.2.2 Particle Image Velocimetry

Mie-Scattering

In contrast to LIF, scattering is a non-resonant process of light-matter interaction at phase boundaries. Scattering intensity is a function of the particle diameter d_p and the incident optical wavelength λ. Based on this, the Mie parameter is defined as [35]:

$$x_{\mathrm{M}} = \frac{\pi d_{\mathrm{p}}}{\lambda}. \tag{2.10}$$

According to x_{M}, three different elastic scattering regimes are characterized in Fig. 2.10. If d_{p} is much larger than λ, i.e., $x_M > 90$, basic laws of geometrical optics apply. In contrast, if d_{p} is much smaller than λ, i.e., $x_M < 1$, Rayleigh scattering occurs with lower scattering intensities. The Mie scattering is observed in the intermediate range, in which the particle diameter is of the same magnitude as the optical wavelength. The scattering intensity reveals sensitive dependence on d_{p}. In the Mie regime, the intensity is approximately proportional to the d_{p}^2, while it turns to be proportional to d_{p}^6 when approaching $x_{\mathrm{M}} = 1$. Figure 2.10 also indicates the angle dependence of the scattering intensity. Forward scattered light (scattering angle $\theta_{\mathrm{s}} = 30°$) has higher intensities than the perpendicularly scattered light ($\theta_{\mathrm{s}} = 90°$).

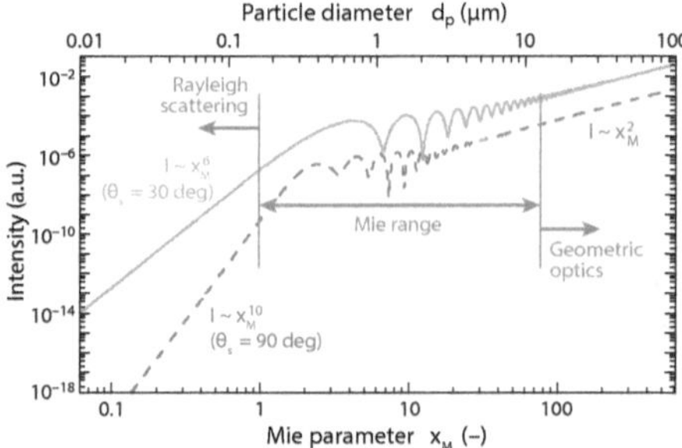

Figure 2.10: Scattered intensity over Mie parameter for two scattering angles θ_s, reprinted from [35].

Figure. 2.11 compares the scattering pattern of differently sized oil droplets with diameters of 1 µm and 10 µm [36], in (a) and (b), respectively. Herein, the same intensity scales apply for both angular patterns. The particle diameter apparently dominates the amplitude of the scattering intensity. More importantly, the scattering intensity shows a very strong variation as a function of the scattering angle θ_{s}. While a sufficient amount of light scatters in the forward direction, the light scattering in the backward direction is weaker. Differences in the scattering intensity require careful consideration when performing stereo-PIV and tomo-PIV using multiple viewing angles.

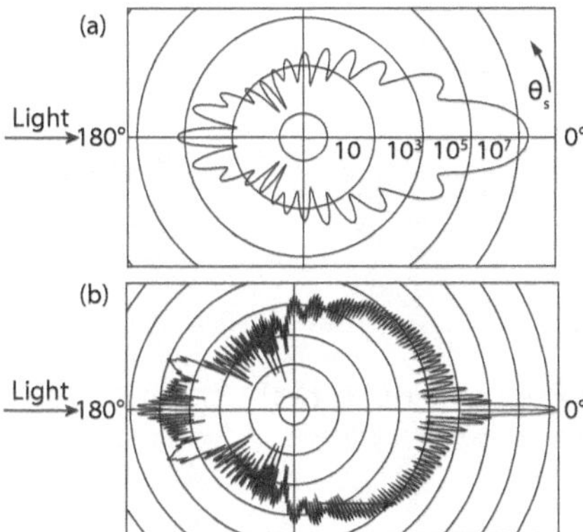

Figure 2.11: Light scattering by (a) 1 µm and (b) a 10 µm oil particle in the air, adapted from [36].

Seeding particles for PIV

In practical PIV measurements, the seeding particles should be sufficiently small to follow the flow motion. Thus, the response behavior of particles needs to be taken into account. With a sudden change in the flow velocity u_f, the slip velocity u_s between particle and fluid is derived from the Stokes' drag law [36]:

$$u_\text{s} = u_\text{p} - u_\text{f} = d_\text{p}^2 \frac{\rho_\text{p} - \rho}{18\mu} a, \tag{2.11}$$

in which u_p and ρ_p are the particle velocity and density respectively, μ is the dynamic viscosity of the fluid, and the particle acceleration a is the temporal variation of the particle velocity, namely, $\mathrm{d}u_\text{p}/\mathrm{d}t$. By assuming $\rho_\text{p} \gg \rho$ in the gas flow, u_p follows an exponential law:

$$u_\text{p} = u_\text{f}[1 - \exp(-\frac{t}{\tau_\text{p}})], \tag{2.12}$$

and the characteristic response time of particle τ_p is derived as: [36]

$$\tau_\text{p} = d_\text{p}^2 \frac{\rho_\text{p}}{18\mu}. \tag{2.13}$$

Therefore, the response time is a function of three dominant parameters: particle diameter, particle density, and fluid viscosity. Usually, the variation of fluid properties and the choice of particle material are restricted by experimental conditions. Hence, the particle diameter should be small enough in order to follow the velocity fluctuations. Basically, particles' response times should be sufficiently shorter than the characteristic time of the relevant gas flow τ_f. To quantify the fidelity of a particle tracer, the particle Stokes number is introduced as

$$Stk = \frac{\tau_\mathrm{p}}{\tau_\mathrm{f}}, \tag{2.14}$$

where τ_f is the characteristic time scale of the flow. For $Stk \ll 1$, particles perfectly follow the flow motion. For practical PIV applications, $Stk < 0.1$ provides a well acceptable tracing accuracy with errors below 1% [37].

Velocity Measurement

Particle image velocimetry is a powerful optical technique for velocity measurements by imaging the Mie scattering of particle tracers seeded in the fluid. In Fig. 2.12, a typical experimental setup of planar, two velocity components (2D-2C) measurements in a wind tunnel is illustrated [36]. The air flow is seeded with small tracer particles, which are illuminated by a double-pulse laser. The beam profile is reshaped to form a laser sheet to allow planar illumination. Particles in the illuminated plane can be imaged twice with a time separation Δt. By comparing the frames at times t_0 and $t_0 + \Delta t$, particle displacement can be evaluated with cross-correlation calculations. To achieve a spatial resolution in 2D, this is commonly accomplished for small segments after subdividing PIV frames into $K \times L$ pixels areas. These segments are termed samples or interrogation windows.

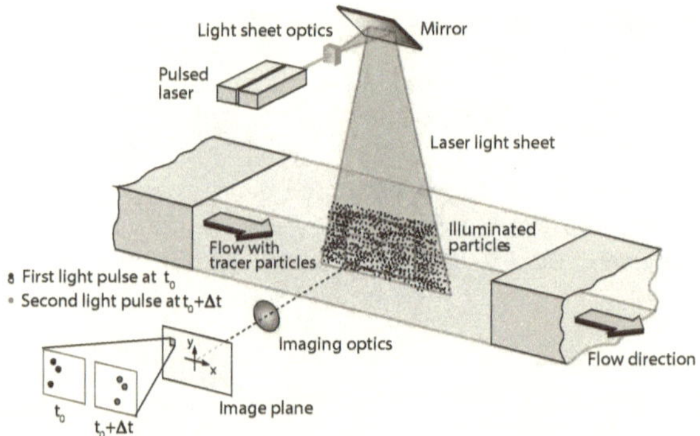

Figure 2.12: Schematic of an experimental setup for 2D-PIV measurement in a wind tunnel, reprinted from [36].

The displacement can be computed in different ways. Among them, a straightforward solution is the discrete calculation of the spatial cross-correlation function [36]:

$$R_{II}(x,y) = \sum_{i=1}^{K} \sum_{j=1}^{L} I(i,j) I'(i + x, j + y), \tag{2.15}$$

where I and I' are intensity values in the interrogation windows of two consecutive frames. I has a window size of $K \times L$ pixels, while I' could take a size larger than that I has. Figure 2.13 illustrates the calculation with an example, in which a smaller I with 32×32 pixels in size is cross-correlated with a larger I' with 64×64 pixels in size. The (i,j) indicates the pixel locations. The (x,y) denotes the possible shift of intensity I which is limited in the range of $-16 < x < +16$ and $-16 < y < +16$. The I is shifted within the I' without crossing over the boundaries. For each shift (x,y), Eq. 2.15 is applied to return a cross-correlation value $R_{II}(x,y)$, which is a sum of the products of all overlapping intensities. With a certain pixel shift (x,y), if the two samples align well, $R_{II}(x,y)$ shows a high value; otherwise, $R_{II}(x,y)$ returns a small value. After performing this for all possible pixel shits, a cross-correlation plane is created with a size of 33×33 pixels. By detecting the highest value (often referred to as the cross-correlation peak), the displacement Δx can be directly estimated which is used to calculate a velocity vector for this interrogation window I.

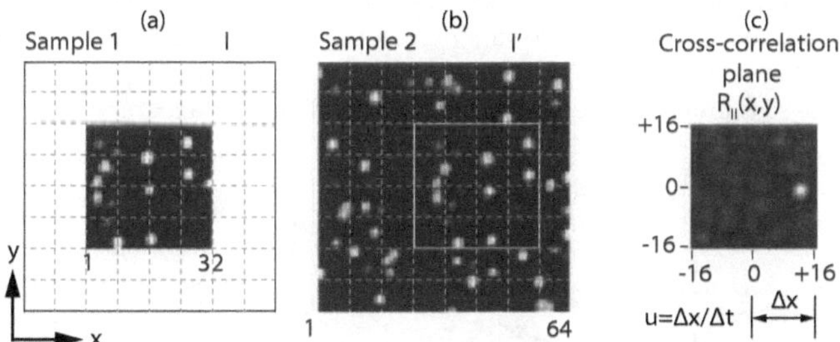

Figure 2.13: The cross-correlation function $R_{II}(x,y)$ in (c) as computed from real data by correlating a smaller sample I (32×32 pixel) in(a) with a larger sample I' (64×64 pixel) in (b) (adapted from [36]).

An obvious disadvantage of the direct cross-correlation calculation is that the computational cost increases proportionally with the size and the number of interrogation windows. This is not efficient to compute PIV recordings with millions of pixels per frame. In practical PIV vector evaluations, instead of applying the direct cross-correlation function, the fast Fourier transformation (FFT) is implemented for interrogation windows, which sufficiently reduces the computation operations. The FFT-based correlation is based on that the cross-correlation of two samples is equivalent to a complex-conjugate multiplication of their Fourier transformations [36]:

$$\hat{R}_{II}(i,j) = \hat{I}(i,j)\hat{I}'(i,j). \tag{2.16}$$

Here, the $\hat{I}(i,j)$ and $\hat{I}'(i,j)$ are the corresponding Fourier transformations of the respective samples $I(i,j)$ and $I'(i,j)$ in Fig. 2.13. Therefore, the summation over all possible pixel shifts in Eq. 2.15 is reduced to a multiplication of the two FFTs. By performing an inverse Fourier transformation on $\hat{R}_{II}(i,j)$, a similar cross-correlation plane as Fig. 2.13(c) can be computed and further utilized for vector calculations after the peak selection.

2.2.3 Laser Scanning

Mirror-based Scanning

For mirror-based scanning, the laser beam is reflected by a moving mirror, and the reflection angles are determined by the periodical changes of the mirror orientations. The commonly used mirror-based mechanical scanners are oscillating mirrors (OM) and polygonal mirrors (PM). The working schemes are schematically depicted in Fig. 2.14. For the OM approach, the mirror is driven by a moving magnet torque motor. The laser beam is scanned back and forth by the periodically oscillating movement of the mirror. The mirror continuously undergoes accelerations and decelerations, thus the scan rates are usually restricted to kHz. This limits its applications in highly turbulent flow, where the scan time period should keep smaller than the characteristic time of flow. In addition, the displacement of scan planes is not uniform due to the unsteady movement. In contrast, the PM is driven by brushless DC motors and runs at a constant rotating speed. The laser beam is scanned only in one direction. Increasing the number of facets or the rotating speed, higher scan rates up to 10 kHz are feasible [38]. Compared to the oscillating mirror, polygonal laser scanning mirrors reveal higher precision and accuracy.

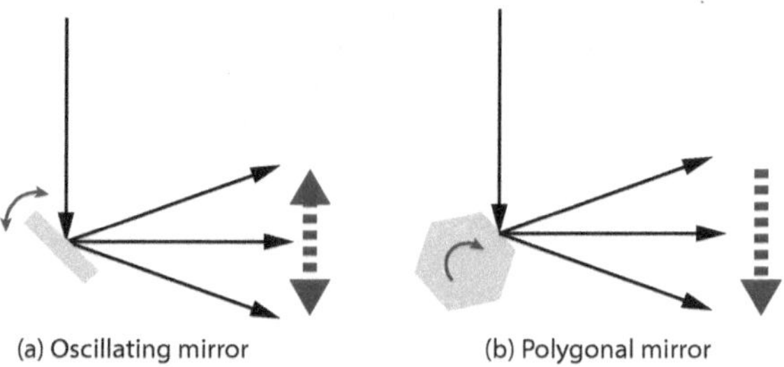

(a) Oscillating mirror   (b) Polygonal mirror

Figure 2.14: Mirror-based mechanical scanners.

AOD-based Scanning

The AOD-based scanning is based on the acousto-optic (AO) diffraction. The AO diffraction can be characterized into two different categories: the Raman-Nath regime and the Bragg regime. The characterization is based on the Raman-Nath parameter Q, which is defined as:

$$Q = \frac{2\pi L \lambda}{\Lambda^2}.$$

$$(2.17)$$

Here, λ and Λ respectively denote the wavelengths of the optical and acoustic beam, and L represents both beams' interaction length. Two diffraction regimes are schematically illustrated in Fig. 2.15. If $Q \ll 1$, the AO interaction occurs in the Raman-Nath regime. The incident laser beam is roughly normal to the acoustic beam, and there are various diffraction orders with intensities defined by the Bessel function (see Fig. 2.15(a)). In contrast, if

$Q \gg 1$, the AO interaction is in the Bragg regime in which only one dominate diffraction order is available (other higher orders exist but with very low intensities), and the zeroth order is the transmission of the incident beam (see Fig. 2.15(b)). At a particular incident angle, the intensity of the first order beam reaches its maximum. For the laser scanning applications, only the Bragg regime is relevant, in which the first order has primary intensity and its diffraction angle θ_{d} changes depending on the acoustic wavelength.

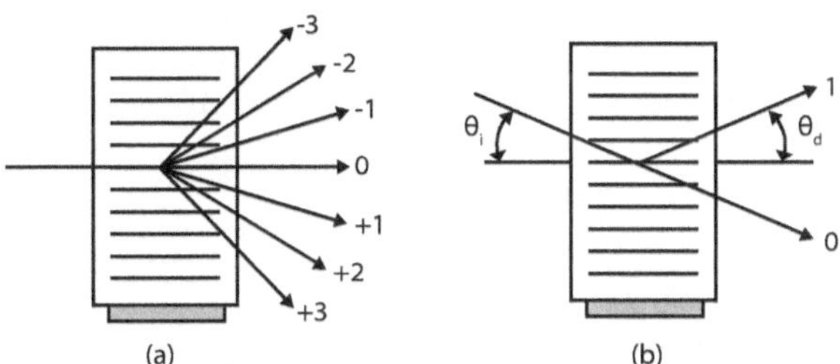

Figure 2.15: Schematics of the Raman-Nath diffraction and the Bragg diffraction.

Generally, there are two common ways to analyze the acousto-optic interaction. The material could be understood as a diffraction grating moving with the sonic velocity. The propagating acoustic wave induces local changes of the material density and the refractive index due to compression and expansion. This leads to periodic changes in the phase of the incident optical wave. Alternatively, the light and sound could be understood as particles, namely photons and phonons, undergoing collision in which the energy and momentum are conserved. Both descriptions can be used to derive the important parameters to understand the acousto-optic diffraction. However, only the latter description is depicted here due to its intuitive nature.

For isotropic acousto-optic diffraction, the refractive indexes n of the medium is the same for the incident and diffracted photons:

$$n = n_{\mathrm{i}} = n_{\mathrm{d}}, \tag{2.18}$$

with n_{i} and n_{d} being the refractive indexes experienced by the incident and diffracted photons. n can be expressed for the optical wave, as:

$$n = \frac{c}{v} = \frac{\lambda_0}{\lambda}, \tag{2.19}$$

where c and v are speeds of light in vacuum and medium, respectively, while the λ_0 and λ are wavelengths of light in vacuum and medium. During the collision, an incident photon absorbs an acoustic phonon's energy and converts to a diffracted photon. The total energy is conserved, which requires:

$$\hbar\omega_{\mathrm{d}} = \hbar\omega_{\mathrm{i}} + \hbar\Omega$$
$$\text{or}$$
$$\omega_{\mathrm{d}} = \omega_{\mathrm{i}} + \Omega. \tag{2.20}$$

Here, the ω_{i}, ω_{d}, and Ω stand for the angular frequencies for the incident photon, the diffracted photon, and the phonon in the order, and $\hbar$ is reduced Planck's constant, $\hbar = h/2\pi$. Because of $\Omega \ll \omega_{\mathrm{d}}, \omega_{\mathrm{i}}$, such that $\omega_{\mathrm{i}} = \omega_{\mathrm{d}}$. The frequency or wavelength of the diffracted photon remains almost the same as that of the incident photon.

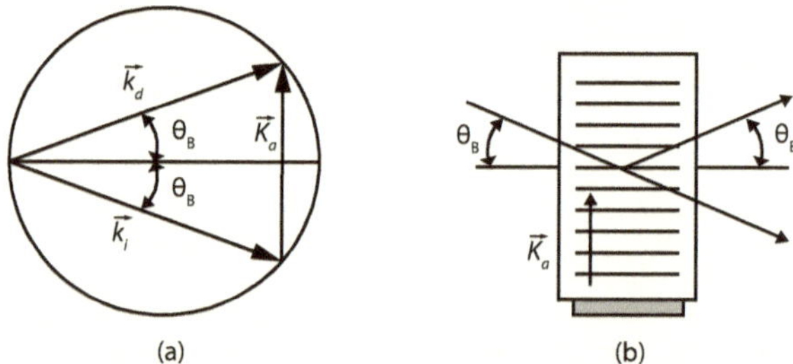

Figure 2.16: Bragg diffraction in an isotropic medium.

Additionally, the momentum conservation during the collision of photon and phonon particles can be expressed by the reconstruction of wave vectors:

$$\vec{k_{\mathrm{d}}} = \vec{k_{\mathrm{i}}} + \vec{K_{\mathrm{a}}}, \tag{2.21}$$

in which $\vec{k_{\mathrm{i}}}$, $\vec{k_{\mathrm{d}}}$, and $\vec{K_{\mathrm{a}}}$ denote the momentum vectors of the incident photon, the diffracted photon, and the phonon in the order. The reconstruction of momentum vectors during collision is illustrated in Fig. 2.16(a). It should be noted that this is only a schematic illustration in which vector lengths do not represent real scales. Amplitudes of momentum vectors are defined as:

$$|\vec{k_{\mathrm{i}}}| = \frac{2\pi n_{\mathrm{i}}}{\lambda_0} = \frac{\omega_{\mathrm{i}} n_{\mathrm{i}}}{c}, \tag{2.22}$$

$$|\vec{k_{\mathrm{d}}}| = |\frac{2\pi n_{\mathrm{d}}}{\lambda_0} = \frac{\omega_{\mathrm{d}} n_{\mathrm{d}}}{c}, \tag{2.23}$$

and

$$|\vec{K_{\mathrm{a}}}| = \frac{2\pi}{\Lambda} = \frac{\Omega}{V}. \tag{2.24}$$

Here, the Ω and V are the angular frequency and the propagation velocity of the acoustic wave in the medium, respectively. With the assumption of $\omega = \omega_{\mathrm{i}} = \omega_{\mathrm{d}}$ and $n = n_{\mathrm{i}} =$

n_d given for the isotropic AO interaction, the amplitudes of the incident and diffracted momentum remain nearly constant, namely, $|\vec{k_d}| = |\vec{k_i}| = 2\pi/\lambda$. As shown in Fig. 2.16(a), the three vectors form an isosceles triangle and the incident angle equals the diffraction angle, which is referred to as Bragg angle θ_B, as highlighted in Fig. 2.16(b). Considering $|\vec{K_a}| \ll |\vec{k}|$ (or $\Lambda \gg \lambda$), θ_B is very small and actually in the order of milliradians. Then, it gives:

$$\begin{aligned}
\theta_B \approx \sin\theta_B &= \frac{1}{2}\frac{|\vec{K_a}|}{|\vec{k}|} \\
&= \frac{1}{2}\frac{\Omega/V}{2\pi/\lambda} \\
&= \frac{\lambda\Omega}{2\pi 2V}.
\end{aligned} \tag{2.25}$$

By applying $\Omega = 2\pi F$ with F being the acoustic frequency, Eq. 2.25 can be simplified to:

$$\theta_B \approx \frac{\lambda F}{2V}. \tag{2.26}$$

The diffracted beam's angle depends on three parameters: the optical wavelength λ, acoustic frequency F, and the acoustic velocity V in the medium. Practically, the λ and F are given by the experimental conditions or devices, such that the Bragg diffraction angle θ_B is nearly proportional to the acoustic frequency F. By periodically varying F, the diffraction angle changes periodically at the same time, and this feature can be utilized to perform laser scanning measurements.

For a certain incident angle θ_i, the acoustic frequency which strictly fulfills the Bragg diffraction is unique. However, in practice, the incident angle of a laser beam can not be adjusted simultaneously with the acoustic frequency. Otherwise, it would require an oscillating movement of the crystal with the same rate as the periodical variation of the acoustic frequency; and it is typically in the order of kHz for applications relevant to the present work. Instead, the incident angle is constant and fulfills the strict Bragg condition for only one certain acoustic frequency. Although the Bragg condition is not strictly satisfied for other alternating frequencies, diffraction still happens but with slightly lower energy flowing into the first diffraction order. For a periodical acoustic frequency changing within a range of (F_{min}, F_{max}), the incident angle could be adjusted to satisfy the strict Bragg condition for the central frequency F_c.

The intensity of the first order beam reaches its maximum I_1 when the Bragg condition is satisfied. The diffraction efficiency η, i.e., the intensity ratio of the diffracted beam and the incident beam, is derived as [39]:

$$\eta = \frac{I_1}{I_0} = \sin^2\left[\frac{\pi^2}{2}\frac{L}{H}M^2\frac{P}{\lambda^2}\right]^{1/2}. \tag{2.27}$$

Here, I_0 is the intensity of the incident beam; P is the acoustic power; L and H are the length and height of the interaction medium. M_2 represents the intrinsic properties of the acousto-optic medium and is denoted as the figure of merit of the material

$$M_2 = (\frac{n^6 p^2}{\rho V^3}), \tag{2.28}$$

where ρ is material density and p is the photoelastic constant. The dominant factors for higher diffraction efficiencies are higher refractive indexes n and lower acoustic velocities V. A larger ratio of L/H is also desired for a higher diffraction efficiency.

Another essential parameter is the rise time T_R of the AO device, which is the response time to a sudden change of the acoustic frequency. Essentially, it represents the acoustic traveling time through the laser beam:

$$T_R = \beta \frac{\phi}{V}, \tag{2.29}$$

where ϕ is the beam diameter, and β is a constant depending on the laser beam profile. In the case of a TEM_{00} beam, β equals 0.66 [39].

2.2.4 Tomographic Reconstruction

The tomographic reconstruction of the intensity distribution in a three-dimensional space by using multiple 2D projections is an essential issue for optical tomography. Several algorithms, including Fourier-domain reconstruction, back-projection, iterative algebraic reconstruction, etc., have been developed [40, 41]. For a limited number of projections in the imaging measurements relevant to this work, the algebraic reconstruction methods are more appropriate [42]. In this regard, a brief introduction is given in the following.

Figure 2.17 illustrates the imaging model with algebraic methods for tomographic reconstruction [42]. The probe volume is often referred to as the *object* space and the projections as the *image* space. The *object* space is discretized in cubic voxel elements with position (X, Y, Z) and intensity $E(X, Y, Z)$. The voxel intensity can be projected along the line-of-sight (LOS) direction onto a camera pixel. For the ith pixel in the *image* space, the (x_i, y_i) and $I(x_i, y_i)$ denote its position and intensity, respectively. A LOS cross-section with a certain bandwidth includes an amount of voxels N_i, contributing to the projected intensity $I(x_i, y_i)$. A weighting factor $w_{i,j}$ gives the jth voxel's contribution to the ith pixel. Thus, $I(x_i, y_i)$ is an intensity summation of all voxels within the cross-section:

$$I(x_i, y_i) = \sum_{j=1}^{j=N_i} w_{i,j} E(X_j, Y_j, Z_j), \tag{2.30}$$

where $E(X_j, Y_j, Z_j)$ is the intensity of the jth voxel. $w_{i,j}$, weights the contribution of voxel intensity $E(X_j, Y_j, Z_j)$ to the pixel intensity $I(x_i, y_i)$. The weighting factor can be calculated as the ratio between the intersecting volume of voxels and the LOS cross-section

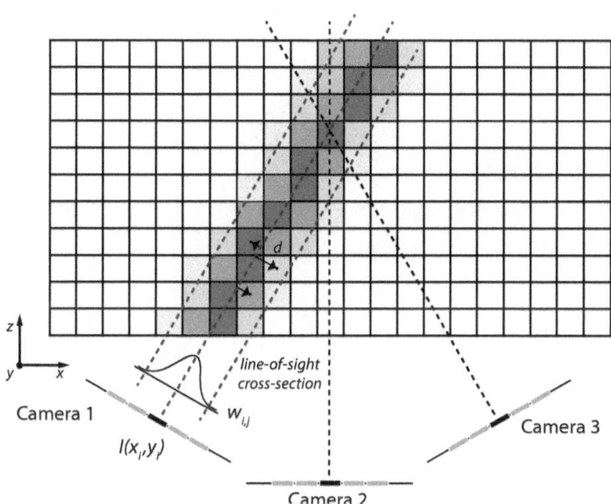

Figure 2.17: Imaging model for the algebraic tomographic reconstruction, adapted from [42].

and the entire cross-section area; thus, it gives $0 \leq w_{i,j} \leq 1$. The value is dependent on the distance d between the voxel center and the LOS, as shown in Fig. 2.17.

To solve this equation, different algorithms have been developed, including the widely used additive algebraic reconstruction techniques (ART or AART) [43] and the multiplicative algebraic reconstruction techniques (MART) [44]. Both methods interactively solve Eq. 2.30 by assuming an initial uniform $E(X,Y,Z)$. The differences are the updated correction between interactive steps, which are expressed as:

$$\mathbf{ART}: E_{k+1}(X_j, Y_j, Z_j) = E_k(X_j, Y_j, Z_j)$$
$$+ \mu w_{i,j} \frac{I(x_i, y_i) - \sum_{j=1}^{j=N_i} w_{i,j} E(X_j, Y_j, Z_j)}{\sum_{j=1}^{j=N_i} w_{i,j}}, \tag{2.31}$$

and

$$\mathbf{MART}: E_{k+1}(X_j, Y_j, Z_j) = E_k(X_j, Y_j, Z_j)$$
$$\cdot \left(\frac{I(x_i, y_i)}{\sum_{j=1}^{j=N_i} w_{i,j} E(X_j, Y_j, Z_j)} \right)^{\mu w_{i,j}}, \tag{2.32}$$

where μ is a scalar relaxation parameter. In the ART, the interactions are updated using a correction defined by the difference of the actual pixel intensity $I(x_i, y_i)$ and the estimated projected intensity $\sum_{j=1}^{j=N_i} w_{i,j} E(X_j, Y_j, Z_j)$. In the MART, corrections are differently determined by a ratio of the actual pixel intensity and the current estimation. In both methods, the weighting factor $w_{i,j}$ guarantees that only the voxels within the cross-section are considered.

In practice, the MART is more suitable than ART for tomographic reconstruction [40, 42] and has been widely used for the volume reconstruction of PIV and LIF recordings [11, 42]. Based the sequential update of the MART (indicated in Eq. 2.32), simultaneous multiplicative algebraic reconstruction techniques (SMART) have been developed to perform more efficient computations [45]. More details about this specific algorithm are not included here.

Chapter 3

Multi-dimensional Laser Diagnostics

In this chapter, studies are performed to address the volumetric scalar and velocity measurements in combustion research using advanced laser diagnostics. First, experimental approaches to accommodate increasing challenges for multi-dimensional imaging measurements are discussed in Section 3.1. Previous studies are summarized within a brief review reporting highlights of individual applications. Then, Section 3.2 presents progressive investigations accomplished within this work divided into three application examples. These investigations focus on different technical aspects but serve the same purpose, namely, improving techniques and methods for 3D optical measurement in combustion. In each part, methods and main results are shortly summarized, and scientific insights are emphasized.

3.1 A Brief Review

Experimental approaches to support laser-based volumetric measurements have been developed for years and can be classified into two categories, namely,

(1) **multiple laser sheets approach**, in which the probe volume is sliced by multiple parallel laser sheets and the time interval between consecutive laser pulses is choosen to be smaller than the characteristic time scale of the flow, and

(2) **volumetric laser illumination**, in which the laser beam is shaped into a collimated laser slab, and multiple cameras from different viewing angles detect projected signals, which are then used for tomographic reconstruction.

In the following, both approaches are briefly introduced with respect to their applications in combustion diagnostics. High-speed flame visualizations are particularly prioritized, and the advantages and drawbacks of diverse technical solutions are included. Because the present work addresses both methods, a brief summary of previous investigations in literature is also provided for each category separately.

Multiple laser sheets approach can be technically realized in different ways such as by using (a) multiple static laser sheets from a laser cluster [46–50], (b) multiple wavelengths generated from a single laser source [51], or (c) a single laser sheet rapidly sweeping through the probe volume. The latter approach is herein referred to as the laser scanning technique. In terms of the laser excitation, the laser scanning can be implemented together with a long-pulse laser [52–54] and multiple camera exposures within a pulse duration, or a high-repetition pulsed laser temporally synchronized with high-speed imaging detectors. The rapid movement of laser sheets can be realized by using mechanical mirrors, i.e., oscillating mirrors [52–59] and rotating polygonal mirrors [38, 60], or by employing acousto-optic deflectors (AOD) [61, 62].

Although mirror-based mechanical scanners have proven their feasibilities in achieving quasi three-dimensional measurements, their maximal scan frequency is restricted by the inertia related to the moving mechanical parts. The mechanical instability remains a major intrinsic limitation, which affects the accuracy and precision of laser beam positions, especially at high scanning speeds. A recently demonstrated alternative to mirror-based scanners is the multiple laser sheet generation using an AOD, which does not contain any moving parts (refer to Section 2.2.3). Therefore, it is advantageous over the mirror-based scanners in terms of scan frequency, accuracy, precision, and spatial resolution.

Table 3.1 provides a brief overview of previous laser scanning measurements in various laminar and turbulent flames. Owing to a broad diversity in technical realizations, only the experiments with kHz-rate scan frequencies f_{scan}, which have immediate relevance to the current work, are included. Pulsed lasers with high repetition rates are mostly required to accommodate scan frequencies in the order of a few kHz, which are necessary to resolve transient processes temporally. A single CMOS camera (coupled with a high-speed image intensifier if necessary) synchronized with the multiple laser sheets is commonly employed to record images from different planes. Compared to the volumetric laser illumination approach, the laser scanning technique enables a better in-plane spatial resolution and fewer constraints on the laser pulse energy and the number of camera views. Among the previous investigations, laboratory-scale applications are found in flow velocity measurements [55, 58, 61, 62], mixture fraction visualization [57] and flame front and reaction zone detection [38, 56, 59, 61].

Moderate scanning frequencies up to 1 kHz can be done by one oscillating mirror [55–57]. To achieve equidistant illumination planes, a constant angular speed of the oscillating mirror is desired. It requires that the scan frequency should be sufficiently lower than the laser pulse repetition rate. At high scan frequencies near 1 kHz, the entire scan sweep is used and scan planes are not equidistant due to the acceleration and deceleration of the mirror. This problem is addressed by using two oscillating mirrors [58, 59], with which both high scan frequencies and nearly constant plane displacement are feasible. An alternative solution is a constantly rotating polygonal mirror, with which the scan rate is increased up to 10 kHz [38]. Combining a fast laser with 50 kHz, PM-based measurements with different spatial and temporal resolutions are achieved to image turbulent flame with $Re \approx$ 5000. More recently, the first AOD-based laser scanning is demonstrated for a turbulent flame [61] and extend to study the in-cylinder flow of an IC engine [62]. In contrast to mechanical mirrors, the plane distances and frequencies are freely adjustable. The AOD scanner used there could support the maximum scan frequency near 1000 kHz without damage or efficiency issues. The extension to a UV laser excitation using AOD is completed in [12, 13, 16] by 3D imaging turbulent gas flames and solid fuel volatile flames. This development is a part of this work and more details will be given in Section 3.2.

Table 3.1: A review of kHz-rate laser scanning measurements in combustion environments.

Authors	Approach	Application	Flow Conditions		Laser	Δz (mm)	Z (mm)	f_{scan} (kHz)
			laminar	turbulent				
Zhang et al. [55]	OM	PIV	-	yes	1 kHz Nd:YAG	1	4	0.2
Cho et al. [56]	OM	OH-LIF	-	yes	5 kHz Nd:YAG pumped dye laser	1.6	8	1
Miller et al. [57]	OM	Toluene-LIF	-	yes	15 kHz Nd:YAG, fourth harmonic	-	10	0.94
Wellander et al. [58]	2 OM	Mie-Scattering	yes	-	20 kHz Nd:YAG	0.79	15	1
Wellander et al. [59]	2 OM	OH-LIF	-	yes	20 kHz Nd:YAG pumped dye laser	0.7	6	0.96
Weinkauff et al. [38]	PM	Mie-Scattering	-	yes	50 kHz Nd:YAG	0.9/2.2/4.2	15	2/5/10
Li et al. [61]	AOD	Mie-Scattering	-	yes	60 kHz Nd:YAG	0.46/.../4.57	13.7	2/.../20
Bode et al. [62]	AOD	PIV	-	yes	9 kHz Nd:YVO$_4$	18	36	3
Li et al. [12]*	AOD	CH$_2$O-LIF	-	yes	100 kHz burst-laser	0.25	2.3	10
Li et al. [15, 16]*	AOD	OH-LIF	yes	-	10 kHz Nd:YAG pumped dye laser	0.42	3.8	1
Zhou et al. [13]*	AOD	CH$_2$O-LIF + TPIV	-	yes	100 kHz burst-laser	0.25	2.3	10

Notes: OM = oscillating mirror; PM = polygonal mirror; AOD = acousto-optic deflector;
Δz = laser sheets discretization; Z = scan depth; f_{scan} = scan frequency.
* A part of the present work

In **volumetric laser illumination**, instead of forming a laser sheet, the beam is expanded and collimated to excite a probe volume. Multiple cameras from diverse viewing angles are employed for simultaneous signal detection. Tomographic reconstructions from the line-of-sight projected signals can be performed using various algorithms, such as ART, MART, and SMART, as introduced in Section. 2.2.4. Volumetric laser illumination can be used for vector and scalar measurements. In terms of the former, tomographic particle image velocimetry (Tomo-PIV) is a well-established technique. For instance, volumetric velocity measurements using Tomo-PIV have been successfully demonstrated in turbulent flames [63, 64] and in-cylinder flows of IC engines [65, 66], to name only a few. A literature review of Tomo-PIV applications is not included here. To measure scalar fields, volumetric flame visualizations can be accomplished with or without laser illumination; the latter is referred to as tomographic chemiluminescence (Tomo-CL). For brevity, only laser-based scalar measurements are summarized in Table 3.2, owning an immediate relevance to the present study. They are single-shot or high-speed, depending on the excitation and detection devices and the concerned physio-chemical process. Demonstrations have been performed on turbulent jet flows using laser-induced fluorescence of iodine (I$_2$-LIF) [67] as well as on laminar and turbulent flames using laser-induced fluorescence of the CH radical (CH-LIF) [68, 69]. In those cases, single-shot tomographic reconstructions were performed using five intensified cameras and a variant of the ART algorithm. Tomographic imaging in turbulent mixing jets was demonstrated using C$_3$H$_6$O-LIF in single-shot [70] and high-speed [71] measurements applying the MART. Additionally, volumetric laser illumination enabled high-speed tomographic OH-LIF [72, 73] and tomographic laser-induced incandescence (LII) measurements [74] of soot particles in turbulent flames. Particular efforts to characterize the spatial resolution were made by comparing tomographic measurements with traditional PLIF techniques [11, 68].

Table 3.2: A review of laser-based tomographic scalar measurements in combustion environments.

Authors	Application	Flow Conditions		Laser	Voxel Size (mm)	Z (mm)	f_{laser} (Hz)
		laminar	turbulent				
Wu et al. [67]	Iodine-LIF	-	yes	Nd:YAG	0.42	50	10
Meyer et al. [74]	LII	-	yes	burst-mode Nd:YAG	0.13	20	10^4
Halls et al. [70]	C_3H_6O-LIF	-	yes	Nd:YAG, fourth harmonic	0.2	15	10
Halls et al. [71]	C_3H_6O-LIF	-	yes	Nd:YAG, fourth harmonic	0.2	35	2×10^4
Halls et al. [75]	CH_2O-PAH-LIF	-	yes	burst-mode Nd:YAG	0.13	30	2×10^4
Ma et al. [68]	CH-LIF		yes	Nd:YAG pumped dye laser	0.18	5	10
Ma et al. [69]	CH-LIF	yes	yes	Nd:YAG pumped dye laser	0.15	32.7	10
Halls et al. [72]	OH-LIF	-	yes	burst-mode Nd:YAG pumped OPO	0.25	15	10^4
Pareja et al. [73]	OH-LIF	-	yes	Nd:YAG pumped dye laser	0.19	27	10^4 (4 pulses)
Li et al. [11]*	OH-LIF	yes	yes	Nd:YAG pumped dye laser	0.075	30	10

Notes: Z = laser volume thickness.
* A part of the present work

3.2 Application Examples

Prior research has thoroughly investigated turbulent flames using two-dimensional optical measurements. However, few research studies have been conducted to address experimental challenges in solid fuel combustion. To extend the volumetric laser diagnostics for the later use in solid fuel combustion, systematic investigations were performed in this work, including both aforementioned tomographic and scanning approaches. These investigations included a feasibility demonstration of tomographic imaging measurements in **paper I**, a turbulent flame topology visualization using high-speed laser scanning in **paper II**, and a study of flame-turbulence interaction with the combined laser scanning and tomographic measurements in **paper III**. While the first two investigations mainly focused on the technical improvement of tomography and scanning separately, the last paper combined both approaches into a simultaneous measurement to understand a complex transient phenomenon. With this strategy, volumetric techniques' potentials were gradually explored, and the limitations of different systems were better understood. Moreover, several scientific questions in turbulent combustion study were addressed, extending knowledge of flame-turbulence interaction. These investigations have been summarized in the first major part of the cumulative results (see Fig. 1.1), namely, *the advancement of volumetric laser diagnostics and applications in turbulent flames*. In the following, the main purposes, primary methods, outcomes, and conclusions of these investigations are summarized.

3.2.1 Tomographic OH-LIF in Laminar and Turbulent Flames

Tomography has been used to study turbulent combustion for years, and for instance, Tomo-PIV is meanwhile a well-known technique to measure spatial structures of turbulent flow. For Tomo-PIV, the particle positions are commonly reconstructed from multiple projections using the aforementioned MART method. In recent years, tomographic techniques for scalar imaging have gained much interest, from the research perspectives, due to the inherent demand to resolve 3D mixing and reaction fields in combustion phenomena. Unlike Tomo-PIV, a higher number of projection views is usually required for an accurate tomographic reconstruction of scalar fields. This requirement becomes even more crucial when switching from laminar to turbulent flames, in which the spatially overlapping structures have to be decoded. It requires more cameras or an upgrade with image doublers in practice, resulting in much fewer pixels available for each projection. Additionally, the flame and mixing visualizations mostly rely on LIF measurements, where the detection wavelength is in the ultraviolet (UV) range. Hence, image intensifiers are needed to convert UV photons to visible photons, leading to an enormous increase in the entire system's complexity and cost. More importantly, the resolution capability of the image intensifier is limited by its micro channel plate (MCP) and phosphor screen, adding another uncertainty in the spatial resolution of the tomographic imaging.

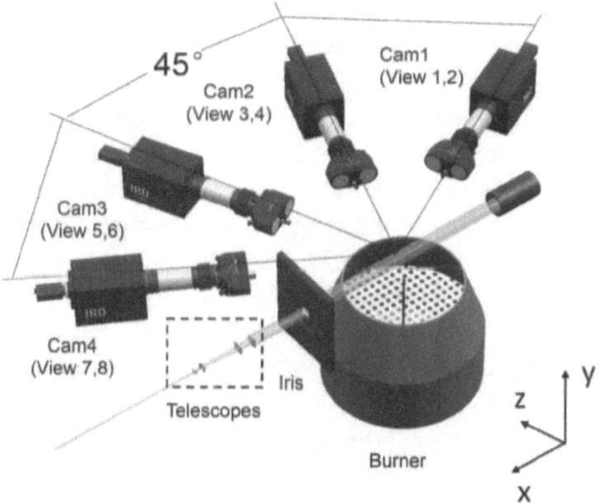

Figure 3.1: Experimental setup for tomographic OH-LIF measurement in a jet-in-coflow burner with four cameras and eight projection views [11].

To address open questions regarding the tomographic capability, volumetric OH-LIF was demonstrated in collaborations with LaVision GmbH (Germany). The purpose of this joint measurement was to examine (1) the effects of the laser slab thickness, projection numbers and angles on the quality of tomographic reconstruction results, (2) the spatial resolution compared to the conventional planar LIF method, and (3) the feasibility to image turbulent flame structures. To fulfill these purposes, a multi-camera detection system of eight projection views was employed for flame structure diagnostics, as shown in Fig. 3.1. Four identical CCD cameras (LaVision, E-lite) respectively equipped with an image intensifier and a UV lens were used to detect the OH-LIF signals, excited by a 10 Hz dye laser system using the $Q_1(8)$ line in the A-X(1-0) transition. To improve the reconstruction accuracy with more perspectives, each imaging set was coupled with an image doubler, which detected the OH distribution from two slightly different angles. Although two projections of an image doubler were overlapped on the CCD sensor, the LIF signals in the reaction zone were mapped onto different pixels, thus that it was feasible to separate them in space. The eight views were distributed over a large angular range of about 150°, which enabled collecting sufficient spatial information of the flame.

Single-shot illumination of the reaction zone was performed firstly in a CH_4/air premixed laminar flame and then extended to visualize a non-premixed turbulent jet flame. For the former, the laser beam with about 25 mJ per pulse was formed into a volume with a cross-section of 30 mm × 30 mm. Figure 3.2(a) shows a sequence of 2D OH-LIF images recorded simultaneously from eight cameras (denoted as views). Based on that, 3D OH fluorescence signals were reconstructed using the SMART approach implemented in Davis 8.4 (LaVision). 150 million voxels discretized the *object* space with a size of 75^3 μm^3. To converge the results and minimize the residual errors, 100 interactive steps were performed for each single-shot reconstruction. The reconstructed flame structure and the top-view intensity summation are shown in Fig. 3.2(b) and (c), respectively.

Taking this configuration as a reference case, two parameter studies were investigated. First, the laser illumination volume thickness was changed with its center remaining aligned with the jet nozzle. Variation in the thickness was achieved by reducing the aperture of a two-dimensional slit. By doing this, the laser fluence was maintained unchanged. The tomographic reconstruction was evaluated with a step-wise reduction of the laser thickness from 30 mm to 10, 4, and 2 mm. The normalized intensity profiles were extracted from the center plane ($z = 0$) and compared with a planar OH-LIF measurement in the same position with a laser thickness of approximately 100 μm. It was found that, with an increasing laser thickness, the intensity curve across the reaction zone became smoother, which indicated a decreasing spatial resolution. It was explained by the less integrated signals in the line-of-sight direction, which could be better resolvable for a thinner illumination volume. The reconstructed OH-LIF results with a 2 mm thickness revealed the best spatial resolution of ∼175 μm. This quality is equivalent to the planar measurement with the much thinner light sheet. With an illumination thickness of 30 mm, the spatial resolution was decreased to approximately 1 mm.

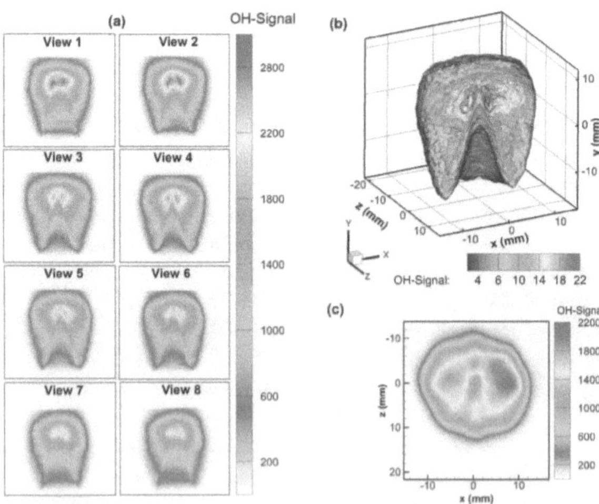

Figure 3.2: (a) OH-LIF images from eight projection views. (b)Tomographic reconstruction of a laminar flame structure using eight projection views. (c) Top-view projections of the reconstructed flame intensities [11].

Second, the effects of the number and angle of projection views were explored. Similar to the first analysis, a laminar flame reconstruction with 30 mm illumination and eight projection views was used as a reference case. By reducing the number of involved projections from 8 to 6 and 4, the quality of reconstruction results slightly decreased in the case of remaining the large angular range of the views. However, if the projection views were constrained within a narrower angular range, reconstructions revealed a dramatic decrease in quality and even artificial signals or structures. This was explained by the inherent similarity of line-of-sight information obtained with closely positioned views, which induced difficulties in properly distributing the pixel intensity into the object space. It can be concluded that a large number of projection views enhance the reconstruction accuracy, and in the case of a limited number of views, a broad range of viewing angles should be guaranteed. More detailed discussions are issued in **paper I** [11].

Moreover, the tomographic OH-LIF was demonstrated in a lifted turbulent jet flame with a non-premixed methane jet. Figure 3.3 shows an instant flame structure visualized by OH-LIF intensity iso-lines. Additionally, the top view and side views of reconstructed flame surfaces are presented in Fig. 3.3 as well. A flame hole structure was observed near the flame base with an estimated diameter of $3 \sim 4$ mm (marked by a yellow circle). Based on the spatial resolution addressed with laminar flame measurements, this structure was considered as fully spatially-resolved. Therefore, it can be concluded that the tomographic OH-LIF measurement is feasible to visualize turbulent flame structures.

The conclusions from this experimental investigation can be summarized as follows. The volumetric flame visualization through tomographic OH-LIF reconstructions is feasible and capable of providing sufficient accuracy and precision compared against well-established OH-PLIF results. This is ensured by the utilization of a large number of projections as well as a broad angular range. In future work, switching from laminar to turbulent conditions, high-speed tomographic measurements are expected to resolve flame evolution temporally. However, high-speed dye lasers' pulse energy is usually limited to a few hundred microjoules. Compared to the low-speed systems with more than ten millijoules per pulse in UV, lower LIF intensities restrict the high-speed applications in turbulent flames. Additionally, as discussed above, multiple CMOS cameras with two-stage image intensifiers are commonly inevitable technical solutions, which are associated with cost-intensive investments and efforts [71–73, 75]. Therefore, a technical alternative to accommodate high-speed laser-based volumetric measurements is needed. This originally motivates a new investigation, which will be discussed in the next section.

3.2.2 Flame Topology with Quasi-4D CH_2O-LIF Laser Scanning

In this section, laser scanning techniques have been developed. To circumvent the issues related to the tomographic flame imaging mentioned above. Instead of using mechanical mirror-based scanners, a novel AOD-based scanning system was established. The first attempt for quasi-4D flame visualization using AOD was demonstrated on a lifted jet flame [61], which was also investigated in Section 3.2.1. To visualize the flame position, oil droplets with a size of a few micrometers were seeded in both fuel jet and co-flow. They underwent evaporation when approaching the high-temperature reaction zone. By detect-

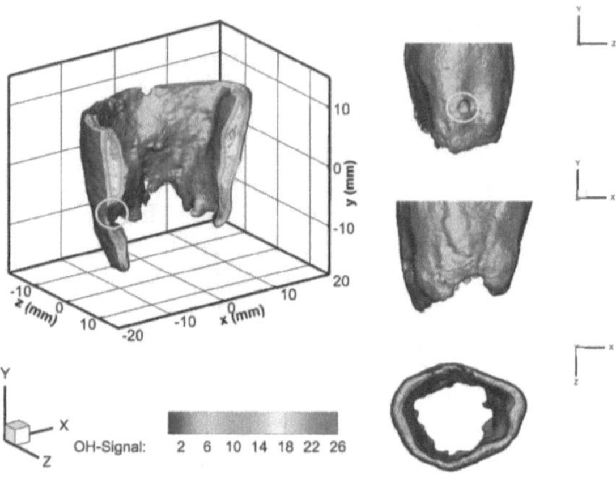

Figure 3.3: Single-shot tomographic reconstruction of a non-premixed lifted turbulent methane jet flame with top view and side views shown in the right column [11].

ing intensity gradients of the Mie scattering, 3D flame structures were reconstructed from multiple 2D images. However, the flame position visualized by oil droplets can not accurately represent the flame reaction zone. A proper flame visualization is only feasible by detecting intermediate species such as OH radicals or CH_2O, which often requires an excitation wavelength in the UV range. Hence an improvement of the AOD-based scanning technique to accommodate the extension from visible to UV excitation is needed.

With this background, experiments were conducted in close collaborations with the Sandia National Laboratories. A volumetric LIF measurement was developed using a high UV-transmission AOD and a pulse-burst laser operated at a repetition rate of 100 kHz. Several perspectives can be summarized for this joint work, including investigations of (1) the feasibility of AOD synchronized with a 100 kHz laser, (2) the capability of laser scanning compared with tomographic approaches to resolve flame structures spatially and temporally, and (3) the flame topology characteristics near the base of a lifted flame. For these purposes, the CH_2O-LIF was used to detect the 3D reaction zone of a partially premixed DME/air lifted turbulent jet flame. The experimental setup is shown in Fig. 3.4, in which a pulse-burst laser at 100 kHz is used for excitation. The 355 nm laser beam was swept through the probe volume with a 10 kHz scan rate resulting in 10 individual scan planes aligned parallel to each other. The laser sheets were parallelized and focused to a thickness of approximately 100 μm. A beam monitor separately registered the spatial discretizations of scan planes in z-direction, and an entire detection volume of approximately $17 \times 12 \times 2.3 \, \text{mm}^3$ was realized. The laser profiles were simultaneously registered with another intensified camera by the LIF imaging of a biacetyl-seeded laminar flow excited with the same laser to account for energy fluctuations and inhomogeneities of the laser sheets.

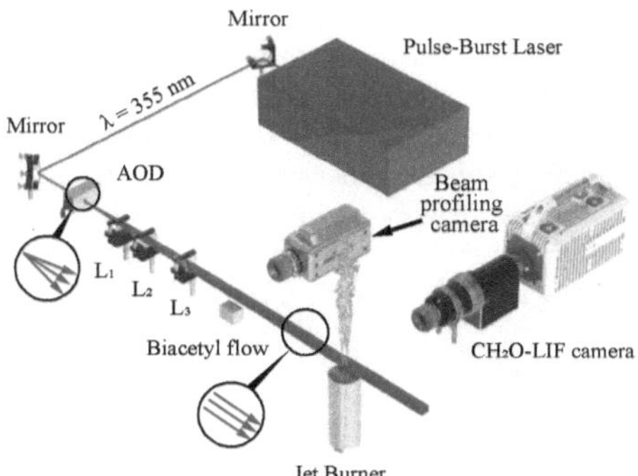

Figure 3.4: Experimental setup for a quasi-4D CH_2O-LIF measurements in a DME/air turbulent lifted jet flame using an AOD combined with a 100 kHz pulse-burst laser [12].

The volumetric reconstruction was performed by linearly interpolating 10 planar images within one scan sequence with a spatial discretization equal to the pixel size. A single-shot reconstruction of volumetric CH_2O-LIF signals is shown in Fig. 3.5(a). The flame structure is represented by the intensity iso-lines of CH_2O-LIF signals. By evaluating the 3D scalar gradients and using an adaptive thresholding method, the 3D surface reaction zone was detected, as shown in Fig. 3.5(b). More details about the processing steps are discussed in **paper II** [12]. Based on the surfaces, flame topology was further characterized by calculating 3D curvatures. Here, the mean curvature κ_m and Gaussian curvature κ_g were derived from the two principal curvatures κ_1 and κ_2. According to the well-known κ_m-κ_g diagram, local shapes were identified for the 3D CH_2O-LIF surfaces.

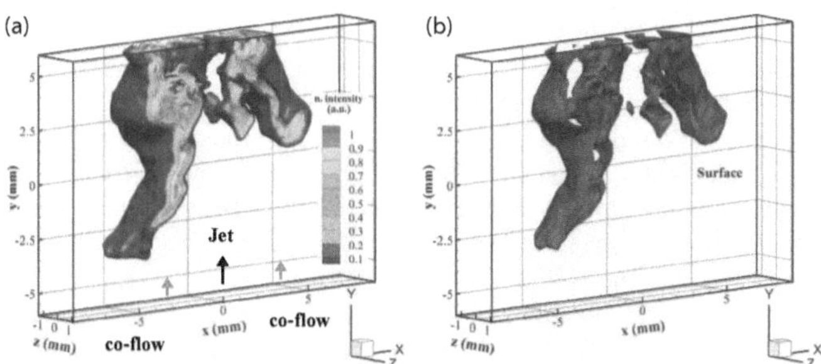

Figure 3.5: (a) Single-shot volumetric reconstruction of CH_2O-LIF signals; (b) 3D surface detection of CH_2O-LIF structures by evaluating the intensity gradients [12].

Based on the curvature analysis, two primary outcomes can be summarized. First, the mean curvature κ_m was compared between 2D and 3D measurements with the same processing methods employed. It was found that the probability density function (PDF) of two-dimensional κ_m was much narrower than the three-dimensional κ_m. It indicated that the lack of out-of-plane information resulted in a significant underestimation of the fully 3D curvatures. Second, the topology characteristics were evaluated by defining a shape factor $sf = \kappa_2/\kappa_1$, which was in a range of [-1,1] due to $|\kappa_2| < |\kappa_1|$. In Fig. 3.6, the shape factors were evaluated for both inner and outer surfaces. For the turbulent flame structures in this study, the cylindrical shapes with $\kappa_m \approx 0$ were preferential, and the probabilities of saddle shapes were higher than that of elliptic shapes. More importantly, the sf PDFs revealed similar trends for inner and outer surfaces, although the curvatures amplitudes of inner surfaces are slightly higher than that of outer surfaces. It suggested that the damping of turbulent fluctuations by the temperature gradients through the pre-heat zone reduced the curvature amplitudes, but the local structure topology remained self-similar.

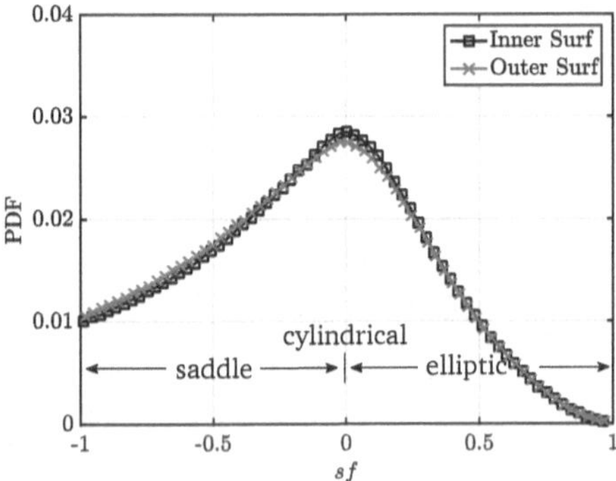

Figure 3.6: PDFs of the shape factor sf for inner and outer CH_2O-LIF surfaces [12].

In terms of diagnostics, statements can be made after comparing the tomographic reconstruction using multiple projection views with laser scanning imaging with an AOD. The laser scanning reveals equivalent spatial resolutions to tomographic methods for applications in laminar and lightly turbulent flames. Sufficient temporal resolution can be realized by synchronizing the scanner with high-speed lasers. Besides, laser scanning requires much lower laser pulse energy than tomography by focusing the laser beam into a sheet instead of forming a volume. More importantly, only one detection camera is needed, which remarkably reduces the imaging system's complexity and cost. The main drawback is the small scan depth restricted by the crystal material used for the UV laser. In contrast, laser-based tomographic scalar imaging is advantageous for applications with a large probe volume. However, numerous issues such as optimization of the spatial resolution, limited pulse energy, and procurement of high-speed intensified cameras make the experiment for studying transient phenomena challenging. Considering the solid fuel combustion study pursued in the present work, the probe volume is restricted to a few millimeters in thickness, but higher spatial resolutions are desired with a limited laser pulse energy. To fulfill these requirements, the AOD-based laser scanning is appropriate for volumetric scalar imaging in this study. And tomographic imaging is implemented to visualize the 3D flow field through Tomo-PIV measurements in future work.

3.2.3 A Study of Flame-Flow Interaction with Combined Approaches

With the knowledge gained by the previous investigations, AOD-based laser scanning and tomographic imaging were combined to perform simultaneous high-speed volumetric scalar and velocity measurements in this section. The experimental work was completed with collaborations with the Sandia National Laboratories. Apart from the diagnostics perspective, a further understanding of the flame stabilization mechanism and the flame-flow interaction was emphasized. The previous quasi-4D flame visualization in [12] was extended with a tomographic PIV measurement and employed to study the DME lifted jet flame. Figure 3.7 shows the experimental setup, in which the methodology of the scanning CH_2O-LIF has been introduced in Section 3.2.2. Additionally, a 10 kHz dual-head Nd:YAG laser illuminated a probe volume with a thickness of approximately 2.3 mm spatially overlapping with the LIF laser sheets. The double PIV pulses with a Δt =15 μs were in the middle of a temporal scan sequence and synchronized with the 100 kHz pulse-burst laser. Four CMOS cameras detected the Mie scattering of 0.3 μm Al_2O_3 particles from different projection angles. The 3D flow fields were assessed by reconstructing the particle location and calculating the spatial cross-correlation to compute vectors. More details are provided in **paper III** [13].

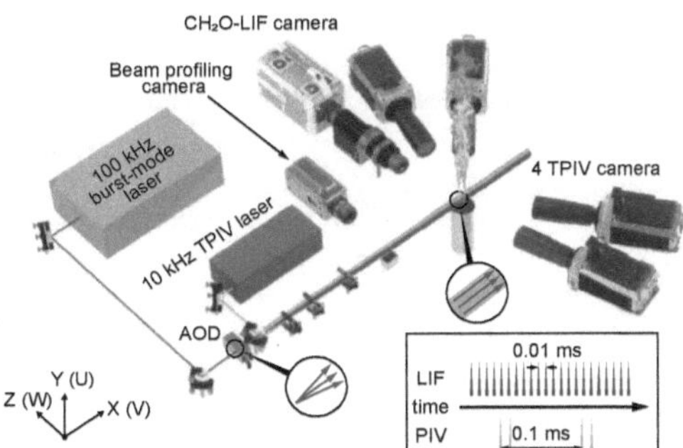

Figure 3.7: Experimental setup of simultaneous CH_2O-LIF and Tomo-PIV measurement in a turbulent lifted jet flame [13].

Based on signal reconstructions from multiple 2D planes, the CH_2O surfaces were detected by applying adaptive thresholding methods for boundary detection [12]. The particle trajectories were evaluated with the Lagrangian analysis by tracking the artificial massless particles (i.e., fluid elements). Figure 3.8 shows particle trajectories together with the inner and outer surfaces. A conditional statistics was performed, in which the trajectories intersected CH_2O inner surfaces at a time instant of $t = 0$ indicated by solid black dots in Fig. 3.8(b). From that time point, the fluid particles were temporally tracked backwards and forwards. Particle trajectories were divided into three zones depending on the initial velocity magnitudes V_{norm}: zone ① for $V_{norm} \leq 0.5$ m/s, zone ② for 0.5 m/s $< V_{norm} \leq 2$ m/s, and zone ③ for $V_{norm} > 2$ m/s. The corresponding trajectories are indicated by different colors in Fig. 3.8, and accompanying velocity components and magnitudes are plotted in Fig. 3.9(a-c). In zone ①, the gas velocity slightly fluctuated around 0.3 m/s when approaching the inner surface, which is then followed by a two-stage acceleration at $t = 0$ and 1.2 ms. This suggested a lean premixed flame behavior illustrated in Fig. 3.9(d). In zone ②, fluid particles accelerated ahead of the inner surface ($t < 0$) from 1 to 4 m/s, implying a strong mixing of air and fuel. No obvious flow acceleration was observed across the surfaces suggesting the absence of strong heat release. This indicated a rich premixed or/and diffusion flame in zone ②. In zone ③, initial gas velocity was significantly high resulted from the high-velocity fuel jet stream. Due to the lack of oxidizer, chemical reactions were suppressed, and the flame could not sustain in this region.

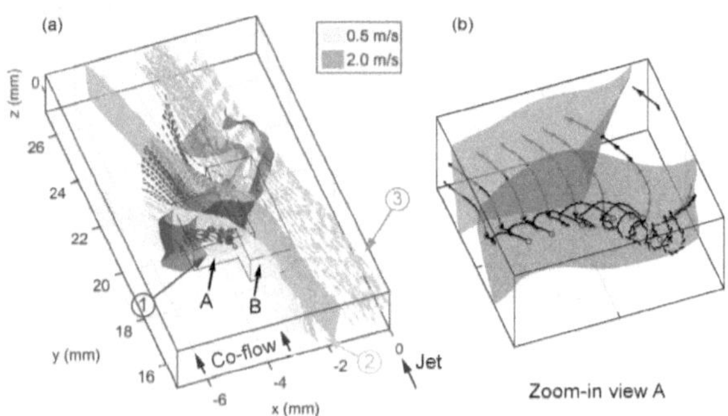

Figure 3.8: (a) Particle trajectories together with the instantaneous CH2O-LIF isosurfaces, veloctiy isocontours correspond to mean velocity magnitude of $V_{norm} = 0.5$ and 2 m/s. (b) A zoom-in view of box A [13].

With this Lagrangian analysis, a hypothesized flame structure with different regimes was proposed in Fig. 3.9(b) based on the observations mentioned above. It suggested that the lean premixed flame near the flame base played an important role in flame stabilization. Additionally, flame displacement speed s_d was evaluated by using the time-resolved gas velocities and flame positions. By correlating s_d with the mean curvature κ_m, the interplay between flame wrinkling and flame propagation was better understood. More discussions are provided in **paper III** [13]. To conclude, with the simultaneous volumetric scalar and vector measurement, a comprehensive study of flame-flow interactions was realized for a deeper understanding of flame stabilization mechanism. Further investigations require additional visualization methods of local mixing processes to improve the verification of local flame regimes.

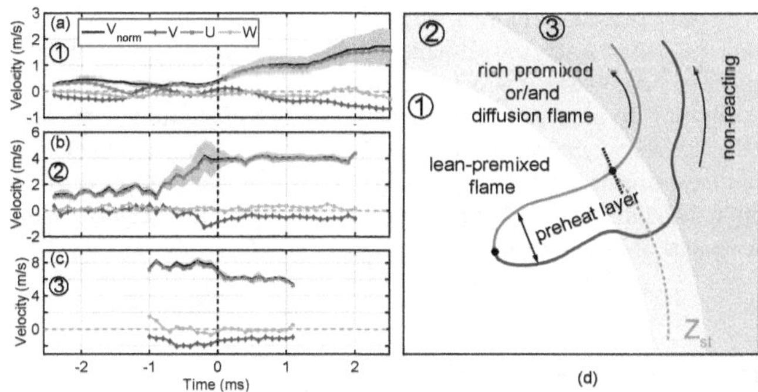

Figure 3.9: Mean velocties along the particle trajectories in (a) zone ①, (b) ② and (c) ③. All trajectories are conditioned at $t = 0$ at which trajectories intersect with the inner CH_2O surfaces. A hypothesized schematic structure corresponding to the CH_2O branch near the flame base in (d) [13].

Chapter 4

Experimental Studies of Solid Fuel Combustion

In the last chapter, the volumetric laser diagnostics have been developed to study transient processes in turbulent flames, and the technical feasibility and capabilities have been comprehensively demonstrated. In this Chapter, these developments are utilized to investigate solid fuel combustion, which is a multi-phase phenomenon involving more complex sub-processes and interactions. First, Section 4.1 briefly summarizes experimental studies in literature concerning optical diagnostics in laminar flow conditions. Solid fuel combustion is then systematically investigated by increasing the complexity from single particles to particle groups and 2D to 3D measurements. Three investigations are introduced, and the main results are summarized in section 4.2.

4.1 State of the art

To study solid fuel combustion at the particle level, drop tube furnaces (DTF) and laminar flow reactors (LFR) are two established configurations to achieve particle heating rates in the order of 10^5 K/s. DTFs are mostly electrically heated, capable of providing controlled wall temperature and gas composition for fuel particle combustion, whereas LFR utilizes combustion products of a hydrocarbon flame to create desired environments. For the latter, Hencken burner, MeKenna-type burner (referred here as flat flame burner, FFB) are different technical realizations. More details regarding the classification of the laminar flow reactor are provided in [21].

In the literature, studies on solid fuel combustion can be divided into two groups, namely single particle combustion (SPC) and particle group combustion (PGC). These two combustion modes differ in how many particles participate in the combustion process or if the particle-particle interaction plays a role. Table 4.1 summarizes several previous experimental investigations giving emphasis on SPC and PGC in laminar flow conditions. For brevity, only in-situ optical measurements with direct relevance to the present work are considered. These measurements are summarized in the following, basically from a diagnostic point of view instead of considering different combustion stages related to the particle residence time. The latter refers to previous discussions in Chapter 2.

To assess the particle surface temperature and the temperature of a volatile-fueled sooty flame, two-color or three-color optical pyrometry is a widely established approach. The temporal history of the particle temperature T_P was used to determine ignition and burnout time [76–78]. This method was optionally combined with optical flame imaging, such as in [79–87], to study single particle combustion processes. The most important disadvantage of pyrometric approaches is that the lowest measurable temperature is limited to around 1200 K. Therefore, the complete temperature rise from the initial heating stages is not detectable, which introduces certain uncertainties in conclusions on the ignition delay time. Additionally, owing to the inherent limitation of line-of-sight measurements, the particle temperature could also be possibly biased by the flame luminescence or the wall thermal radiation.

High-speed (HS) imaging at kHz rates with CMOS cameras provides a more intuitive impression of particle combustion's temporal evolution. In the literature, it was usually implemented to detect the visible light emission without spectral filtering [77, 79–85, 88–90]. Although it has been used to determine the ignition time, the accuracy remains questionable due to the non-sensitivity of the CMOS camera to CH* and OH* chemiluminescence in the UV range. HS flame imaging was performed with [91, 92] and without [77, 79–85] diffusive backlight-illumination (DBI). Herein, backlight enabled the particle position detection prior to ignition, but not resolving particle size and shape (low spatial resolution) with exceptions reported in [10, 15, 16]. Nevertheless, previous investigations successfully demonstrated the capability of detecting particle trajectories [89, 90], flame shape [91–93], and fragmentation [88], improving the understanding of solid fuel combustion fundamentals.

Intermediate species, such as OH or CH radicals, have been seen as useful indicators to visualize gas-phase hydrocarbon flames, and the corresponding experimental access to these species is also essential to study solid fuel combustion. Time-averaged CH* chemiluminescence was imaged by an intensified CCD camera and used to evaluated the particle ignition and volatile combustion time [94]. Recently, Köser et al. [8] applied high-speed OH-PLIF to visualize the reaction zone of single-particle volatile flames. Further investigations included highly resolved DBI and luminescence (LU) imaging with a band-pass spectral filter allowed for the simultaneous data acquisition of ignition delay time t_{ign}, volatile combustion time t_{vol}, particle size and motion with sufficient temporal and spatial resolutions [9, 10, 21]. A high imaging magnification was realized herein by coupling a CMOS camera with a long-distance microscope. Other attempts to spatially resolve single particles and their associated sooty flames were reported in [95, 96].

Regarding particle group combustion, Table 4.1 indicates a considerable diversity in the experimental apparatus and analysis approaches. Cost-effective digital single-lens reflex (DLSR) cameras were adapted to record color images of burning particle groups [97] or perform pyrometric measurements using the RGB channels [99, 101]. Temperature information was obtainable via pyrometry measurements [97, 99, 101], whereas a clear distinction between surface temperature and soot temperature remained not feasible, especially for bituminous coal. Similar to the efforts made to study SPC, imaging approaches are diverse. CH* chemiluminescence imaging using spectral filter [98, 99], broad-band thermal radiation detection using ICCD [100, 101], or CMOS cameras [103–105] were reported to

Table 4.1: In-situ optical measurements of solid fuel particle combustion in laminar conditions with high heating rates.

Authors	Burner	Optical Approaches	Fuel	Mode	Atmospheres	Variations
Timothy et al. [76, 77]	DTF	Pyrometry/HS imaging [77]	3 bit, 2 lig	SPC	O_2/Ar, O_2/He	d_P, O_2, T_{wall}
Levendis et al. [78–84]	DTF	Pyrometry/HS imaging	hvb, sub-bit, lig, biomass	SPC	Air, oxy-fuel	d_P, O_2
Zou et al. [85]	DTF	Pyrometry/HS imaging	bit	SPC	O_2/H_2O	O_2
Tahmasebi et al. [88]	DTF	HS imaging	lig	SPC	Air	moisture
Kim et al. [93]	Hencken	DLSR/LS imaging (ICCD)	sub-bit	SPC	Air, oxy-fuel	O_2
Molina et al. [94]	Hencken	CH* imaging	hvb	SPC	Air, oxy-fuel	O_2
Shaddix et al. [95]	Hencken	Soot imaging (LII)	hvb, sub-bit	SPC	Air, oxy-fuel	O_2
Adeosun et al. [91]	Hencken	HS DBI (low resolution)	sub-bit	SPC	Air (R-to-O)	T_{gas}
Vorobiev et al. [86, 87]	Hencken	Pyrometry/LS DBI	hvb, biomass	SPC	Oxy-fuel	O_2, H_2O
Yao et al. [96]	Mekenna	Digital in-line holography	bit	SPC	Air	-
Köser et al. [8, 9]	LFR	HS OH-PLIF	hvb	SPC	Air, oxy-fuel	O_2
Köser et al. [10]	LFR	HS OH-PLIF/DBI/LU	hvb	SPC	Air, oxy-fuel	d_P, O_2
Li et al. [14]*	LFR	HS OH-PLIF/DBI/LU	hvb	SPC	Air, oxy-fuel	d_P, O_2
Lee et al. [89, 90]	JIC	HS imaging	bit	SPC	Air/O_2	d_P, O_2
Mock et al. [92]	JIC	HS DBI (low resolution)	sub-bit, wood, blends	SPC	Air/O_2	d_P, O_2
Zeng et al. [97]	Hencken	Pyrometry(Spectrometer)/DLSR	sub-bit	PGC	Air, oxy-fuel	O_2
Liu et al. [98]	Hencken	CH* imaging	hvb/sub-bit	PGC	Air, oxy-fuel	O_2
Yuan et al. [99]	Hencken	CH* imaging/Pyrometry(DLSR)	lig	PGC	Air	T_{gas}, O_2
Yuan et al. [100]	Hencken	LS imaging (ICCD)	lig/bit	PGC	Air	T_{gas}, O_2
Yuan et al. [101]	Hencken	LS imaging (ICCD)/Pyrometry(DLSR)	lig/char	PGC	Air, oxy-fuel	T_{gas}, O_2
Xu et al. [102]	Hencken	LS OH-PLIF	bit	PGC	Air	Re_{jet}
Adeosun et al. [103]	Hencken	HS DBI (low resolution)	sub-bit	PGC	Air (R-to-O)	O_2
Sarroza et al. [104]	DTF	HS imaging	anth/bit/lig/biomass	PGC	Air	Co-firing
Prationo et al. [105]	McKenna	HS imaging	lig/sub-bit/bit	PGC	Air, oxy-fuel	O_2, H_2O_{vap}
Li et al. [15, 16]*	LFR	HS 3D OH-LIF/DBI	hvb	PGC	Air	PND, d_P

Notes: bit = bituminous coal; hvb = high-volatile bituminous coal; lig = lignite coal; anth = anthracite; JIC = Jet-in-crossflow; HS =high-speed; LS =low-speed; DBI = diffuse backlight-illumination; LU = Luminescence; DLSR = digital single-lens reflex; SPC = single particle combustion; GPC = group particle combustion; R-to-O = reducing-to-oxidizing
* A part of the present work

investigate particle group combustion.

With respect to data analyses, previous studies mostly focused on the characterization of ignition time or height of particle groups. The imaging techniques listed in Table 4.1 for PGC are all below the spatial resolution that would be required to resolve single particles. The evaluation was therefore based on the temporal integration of the particle intensity into a luminous streak. For instance, the ignition time was determined based on time-averaged high-speed images when the luminescence intensity fulfilled a specific criterion. It has been defined as 12% of the maximum intensity [104], the maximum intensity gradient position [103], and the time point of the first appearance [105]. Similarly, a 10% threshold of the intensity peak was selected in [97, 100, 101] to characterize the ignition delay time from a temporal integration of recorded images, whereas a 50% was chosen in [98] to give better insights into particle group ignition processes. More recent advancements can be found in single-shot OH-LIF measurements of dispersed coal particle streams in a Hencken burner [102]. In that study, the ignition and volatile combustion time were defined by the maximal slope of averaged OH-LIF intensity profiles. It should be noticed that due to the large diversity in methodology reported in the literature, results and conclusions may largely depend on the selected measurement techniques and their uncertainties.

4.2 Application Examples

To understand the individual sub-processes and their interactions of the multi-phase phenomena, experimental efforts were made in the present work focusing on the coal particle combustion in air and oxy-fuel conditions. To evaluate individual parametric sensitivities, experiments were performed in a generic laminar flow reactor, emphasizing on the volatile combustion with high particle heating rates. The previously introduced multi-parameter and multi-dimensional optical diagnostics (in Section 3) are adapted herein to decode solid fuel combustion complexity. Highly resolved optical experiments aim to improve data quality in terms of resolution and accuracy, and minimize the uncertainties in analysis and interpretation. With a step-wise increase in complexity, experimental attempts started with multi-parameter 2D measurements in single particle combustion and were extended to multi-parameter 3D measurements in particle group combustion. Theses investigations are the second part of the main results (refer to Fig. 1.1), namely *the application of advanced optical diagnostics in solid fuel combustion*. The scientific endeavor was condensed in three publications, **paper IV**, **V**, and **VI**, which are briefly summarized in the following sections.

4.2.1 Multi-parameter Measurements of Single Particle Combustion

As pointed out in Section 4.1, the experimental methods to determine the ignition delay time t_{ign} and volatile combustion duration t_{vol} are ambiguous. Especially for high particle heating rates, these parameters are essential to understand fundamental aspects e.g., flame stabilization and particle-flow interaction in oxy-fuel combustion. For a better quantification of t_{ign} and t_{vol}, multi-parameter optical measurements were performed for single particle combustion in a generic laminar flow reactor shown in Fig. 4.1(a). A premixed CH_4 flame was stabilized above a ceramic matrix to generate high-temperature environments, in which bituminous coal particles were injected through a central tube. Different inlet gas compositions were realized to create conventional and oxy-fuel atmospheres. The boundary conditions were characterized by PIV and quantitative OH-LIF measurements. Ignition and volatile combustion were visualized and analyzed with variation in inlet parameters such as O_2 concentration and fuel particle size. To address undergoing pyhsico-chemical sub-processes, multiple imaging techniques including high-speed OH-PLIF, LU imaging, and DBI were employed simultaneously, which are shown in Fig. 4.1(b). The OH-PLIF with a relatively large thickness of 1 mm was used to detect the onset of ignition. The LU intensities aimed to detect the end of gas-phase flames. Moreover, the DBI system with a high spatial resolution realized by a long-distance microscope was used to determine the particle size. All systems were operated simultaneously at 10 kHz. More details about the experimental methodology are presented in **paper IV** [14]. In the following, some essential results are selected and presented here.

Figure 4.2(a-c) respectively show single-shot images from OH-PLIF, LU, and DBI measurements. By temporally tracking the OH-LIF intensity, the ignition delay time was detected by a signal and structure (SAS) analysis within a ROI centered at the particle

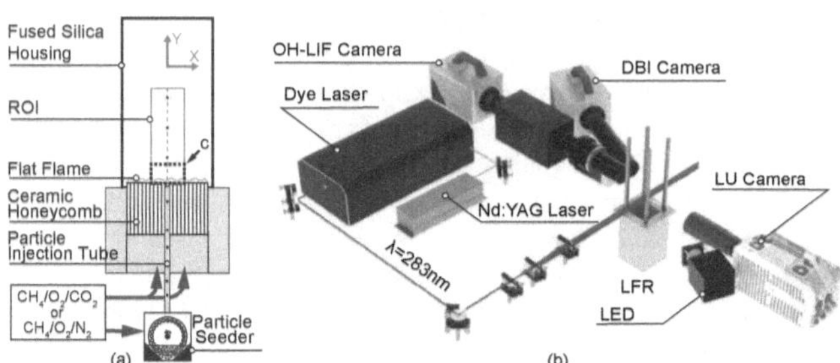

Figure 4.1: (a) Schematic of the laminar flow reactor. (b) Experimental setup of multi-parameter experiments including OH-PLIF, luminescence imaging, and diffuse backlight-illumination. Adapted from [14].

centroid (dashed rectangle in Fig. 4.2(a)). Compared to the luminescence imaging, OH-LIF revealed higher accuracy for detecting ignition due to less sensitivity to the composition of released volatile matter. The luminescence signals were indicators for soot and char combustion and thus used to determine the end time of volatile oxidation $t_{\mathrm{vol,end}}$ based on the temporal variation of the signal area A_{LU}, as shown in Fig. 4.2(d). Different stages of single particle combustion can be characterized by combining simultaneously acquired data. The volatile combustion duration was then calculated by combining information from two measurements: $t_{\mathrm{vol}} = t_{\mathrm{vol,end}} - t_{\mathrm{ign}}$. Moreover, simultaneous DBI measurements allowed for in-situ particle size detection by evaluating the circle-equivalent diameter d_{p}. On this basis, conditional statistics of t_{ign} and t_{vol} were performed for different atmospheres and particle sizes.

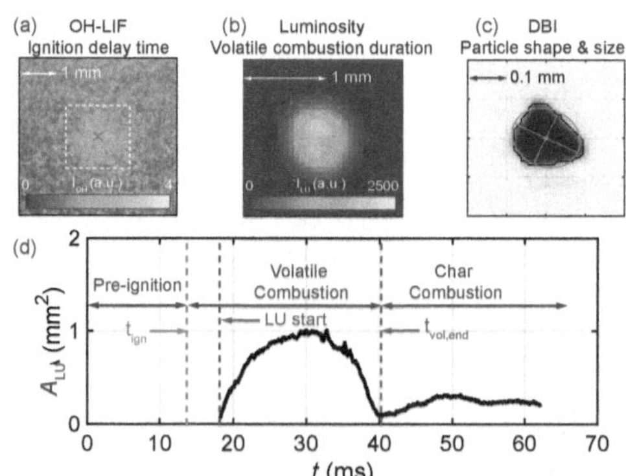

Figure 4.2: (a-c) Individual single-shot visualization of OH-LIF, luminescence, and DBI measurements. (d) Temporal variation of luminescence signal area A_{LU} of a single particle and the characterization of combustion stages. Modified from [14].

In Fig. 4.3, t_{ign} and t_{vol} are shown for two coal samples, A and B with mean diameters of 120 µm and 205 µm, respectively, for varying O_2 mole fractions in N_2 and CO_2 conditions. Important observations can be concluded as follows. (1) Both t_{ign} and t_{vol} evidently increase with increasing particle diameter. (2) For larger particles B, t_{ign} and t_{vol} decrease remarkably with increasing oxygen concentration. This tendency is still observed for smaller particles A, whereas it is less sensitive to the oxygen increase than that of larger particles. (3) With N_2 replacement by CO_2, t_{vol} increases, especially for larger particles. However, the ignition time t_{ign} is almost not affected for smaller particles, and even a slightly lower t_{ign} is noticed in CO_2 conditions for larger particles.

Based on the experimental configurations, numerical simulations were performed within a collaboration with the RWTH Aachen University. The simulations were first validated against experiments and then used to provide further insights into gas mixtures and temperatures. Combining experimental and numerical results, the effects of O_2 mole fraction, particle size, slip velocity, and CO_2 were analyzed and discussed. Herein, special efforts were made to understand the devolatilization and gas ignition process with N_2 replacement by CO_2, considering the change in gas properties such as heat capacity, conductivity, and diffusivity. It was found that, with the same amount of heat extracted from the gas phase, the local gas temperature decreased less in CO_2 than in N_2 atmospheres. Thus, the volatile release was promoted from the less significant cooling effect in CO_2 atmospheres, which accelerated the flammable mixture formation. This process compensated the negative impacts on ignition delay time introduced by the higher heat capacity of CO_2 and probably resulted in an early ignition. It was emphasized that the local heating conditions and gas mixture fraction were dominating factors for understanding the local gas ignition phenomenon and needed to be considered carefully. More details are discussed in **paper IV** [14].

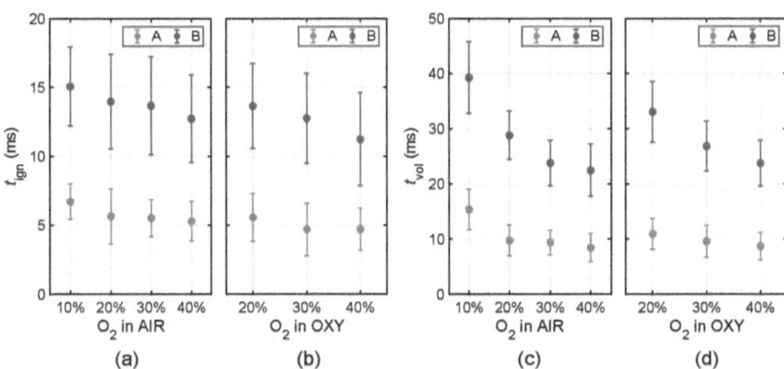

Figure 4.3: Ignition delay time t_{ign} for two particle sizes A and B in (a) air and (b) oxy-fuel conditions. Volatile combustion duration t_{vol} for two particle sizes A and B in (c) air and (d) oxy-fuel conditions. Adapted from [14].

In conclusion, the LFR provided a unique platform with defined boundary conditions and operating flexibility for experimental studies on solid fuel combustion. The ignition and combustion behavior of single coal particles were comprehensively investigated in N_2 and CO_2 atmospheres. Multi-parameter optical diagnostics were essential to analyze the undergoing processes, and physical processes were further explored by combining the results from simulations. This joint experimental-numerical investigation presented a novel approach to understand the fundamentals in oxy-fuel combustion in laminar conditions. On this basis, a deeper understanding of the volatile combustion of isolated single particles was obtained by discussing the physico-chemical effects of various process parameters. For further analysis, 3D laser diagnostics are employed for single particle combustion in the next section.

4.2.2 Volumetric Measurements of Single Particle Combustion

Switching from single particle to particle group combustion, the inherent 3D nature of flame structures must be considered for the volatile flame visualization. Based on the progress made in the last section, the 2D measurements were extended to volumetric imaging measurements by implementing the AOD-based laser scanning technique introduced in Chapter 3. With a step-wise increase of complexity, the volumetric imaging methods were first demonstrated to emphasize single particle combustion in this section and extended to study particle group combustion in Section 4.2.3. The first attempt of volumetric imaging on the single-particle level summarized in **paper V** is briefly introduced in the following.

Similar to the study in Section 4.2.1, the volatile combustion of Colombian high-volatile coal was experimentally studied and a conventional air atmosphere with 10% oxygen was used. Herein, quasi-4D OH-LIF measurements were performed in the LFR to visualize the 3D shape of volatile flames. Igniting particles were temporally and spatially resolved by profiling 3D OH-LIF intensities in the gas-phase flame. Figure 4.4 illustrates the experimental setup for the simultaneous DBI and quasi-4D OH-LIF measurements. The aforementioned AOD-based laser scanning was employed with a probe volume of about 4 mm in depth. Planar OH-LIF images of single-particle volatile flame were detected at different z positions with a high-speed image intensifier camera. Entirely, ten scan planes with a scan frequency of 1 kHz were used to reconstruct the OH-LIF intensity distribution in 3D space. A DBI system, which was slightly angled to the OH-LIF camera, was operated at 10 kHz to address the instantaneous particle size and shape.

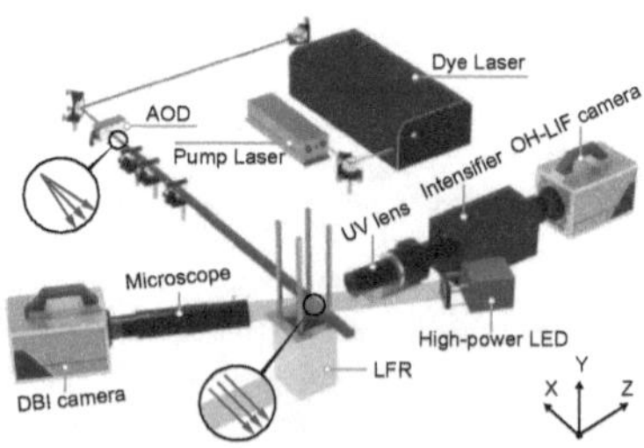

Figure 4.4: Experimental setup of quasi-4D OH-LIF and DBI measurements on the laminar flow reactor. Modified from **paper V** [15].

Figure 4.5(a-c) show examples of 2D OH-LIF images in different z positions within one scan sequence. At $z = -1.04$ mm, a long dark streak in the negative side of the x-axis indicated that the particle was approximately located centrally within the laser sheet. Correspondingly, the laser illumination plane was located behind and before the particle in (a) and (b), respectively. With a linear interpolation of the entire ten planes per scan, a volatile flame structure was reconstructed, shown in Fig. 4.5(d) with intensity iso-contours. The flame structure revealed approximately a symmetric shape, whereas it extended wider above the particle due to the slip velocity between the gas and particle.

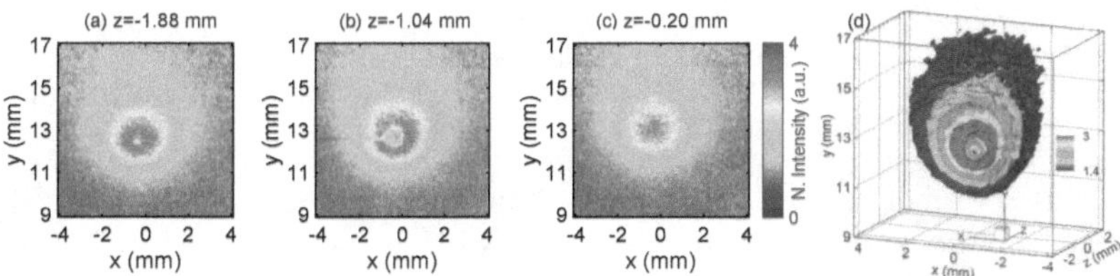

Figure 4.5: (a-c) Three consecutive 2D OH-LIF images at different z-positions. (d) 3D reconstruction of single-particle volatile flame. Adapted from [15].

Radial intensity profiles were extracted from individual reconstructions and were temporally tracked to detect the onset of ignition. DBI measurements provided not only the in-situ particle size but also the particle center to trace particle positions. By applying a particular threshold referenced to the background, the onset of ignition could be precisely determined. The temporal and spatial evolution of global flame structures was then further visualized, starting from the gas-phase ignition. A detailed discussion is provided in **paper V**.

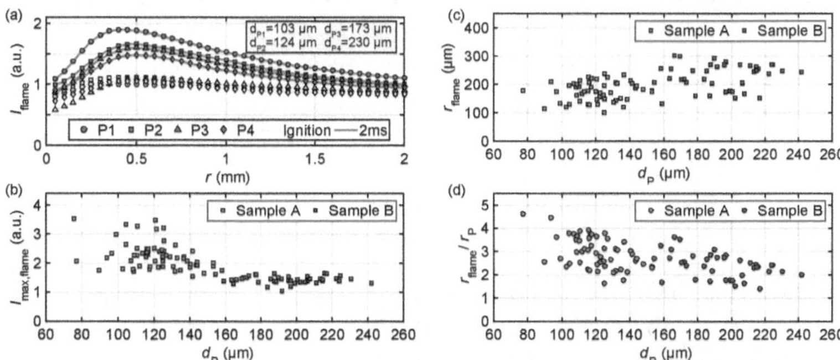

Figure 4.6: (a) Radial OH-LIF intensity profiles of four different single particle volatile flame at ignition and 2 ms after ignition. (b) Maximum flame OH-LIF intensity $I_{\mathrm{max,flame}}$, (c) flame stand-off distance r_{flame} and (d) dimensionless flame distance $r_{\mathrm{flame}}/r_{\mathrm{p}}$ over d_{p} at 2 ms after the onset of ignition [15].

Combining multiple parameters from quasi-4D OH-LIF and DBI measurements, the effect of particle size on the flame topology was explored. Figure 4.6(a) shows the radial intensity I_{flame} of four individual particles with an increasing equivalent particle diameter d_{p} at the onset of ignition and 2 ms after ignition. While their intensities were similar for different sizes at ignition, smaller particles revealed higher I_{flame} at 2 ms. Again, two samples A and B, with different particle size distributions, were used for statistical analysis. By evaluating the maximum intensity $I_{\mathrm{max,flame}}$ at 2 ms, a non-linear decrease of $I_{\mathrm{max,flame}}$ with increasing d_{p} was observed in Figure 4.6(b). It indicated a drop of flame intensity due to higher energy transfer required from the gas to solid phase for particle heating. In addition, Figure 4.6(c) shows the flame stand-off distance r_{flame} calculated based on the maximum gradients of I_{flame}. It implied that the peak heat release position increased with d_{p} due to the higher volatile mass released from larger particles. However, the dimensionless flame stand-off distance $r_{\text{flame}}/r_{\text{p}}$ decreased with d_{p}, as shown in Fig. 4.6(d). It was explained by the lower heating rates and slower volatile release processes of larger particles. Overall, the dimensionless distances varied between $2r_{\text{p}}$ and $4r_{\text{p}}$ for the bituminous coal in this work.

Moreover, the diagnostic feasibility for understanding group particle combustion was demonstrated with different particle number densities, and an in-depth discussion considering the literature was also devoted (see paper **V**). To summarize, this work presents a novel high-speed volumetric optical measurement in the solid fuel combustion research. A fully characterized AOD-scanning method developed in recent years allows for understanding the fundamental particle-flame interaction. With the diagnostics feasibility validated, investigations on single particle combustion are extended to measure group particle volatile flames in the next section.

4.2.3 Volumetric and Multi-parameter Measurements of Group Particle Combustion

In this section, the transition from single particle to particle group combustion is investigated with the similar experimental methodology reported in [15]. Hence, a description of the operating conditions and optical detection systems is omitted for brevity. The analysis methods and main outcomes of **paper VI** are summarized as follows.

One of the major problems for particle group combustion studies is the quantification of the particle seeding density. Two parameters were evaluated based on binary DBI images to address this issue, namely the particle number density PND and the interaction factor IF. Figure 4.7(a) shows an instantaneous binary DBI image with boundaries (dashed lines) indicating the location of the most particles during the entire measurement. A 3D particle jet volume can be generated from the boundaries by assuming symmetry, in Fig. 4.7(b). Then, PND was defined as the ratio of the number of particles N to the volume V: PND $= N/V$. The values of PND can be evaluated for a freely selected particle volume. To calculate IF, for every individual particle with size d_{p}, the nearest neighboring particle with size d_{p1} was detected in the first step, as shown in Fig. 4.7(c). In the second step, the distance between these two particles $L_{\text{pp,1}}$ was calculated. With that, the interaction was characterized by IF $= 0.5(d_{\text{p}} + d_{\text{p1}})/l_{\text{pp,1}}$. Fig. 4.7(d) shows an instantaneous IF map, for which the IF values were interpolated in the area where no particles exist. It was reported that both parameters provided a quantified measure for the in-situ seeding density in particle groups.

With similar reconstruction algorithms proposed in **paper V**, the volatile flame structure can be evaluated with the scanning OH-LIF for particle group combustion. Figure 4.8 exemplifies four individual volatile flames with OH-LIF iso-contours, denoted as case A, B, C, and D. Within the particle jet (in gray), the PND increases from 0.18 to 1.9/mm^3. Several observations can be summarized during this transition: (1) the flame base becomes

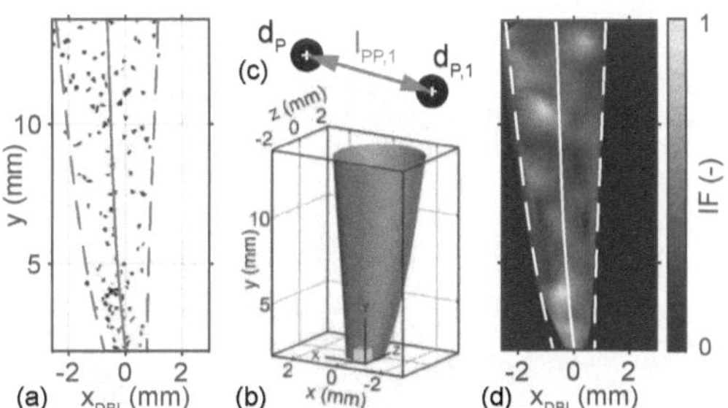

Figure 4.7: (a) A binary DBI image with particle boundaries. (b) 3D particle jet volume. (c) Scheme for interaction factor calculation. (d) An interaction factor map [16].

higher with increasing PND indicating a delay of ignition; (2) the flame structure extends wider presumably due to the increasing volatile mass; and (3) a non-flammable region is formed inside the flame and enlarges at higher PNDs. To understand the transition, the gas-temperature T_{gas} before ignition was estimated based on the OH-LIF intensity profiles. It was found that T_{gas} decreased with increasing PND, which was explained by the enhanced heat transfer between gas and particles. The low temperature and rich gas mixture were considered as the major reasons for flame extinction, which has been further investigated in another study by including numerical simulations [106].

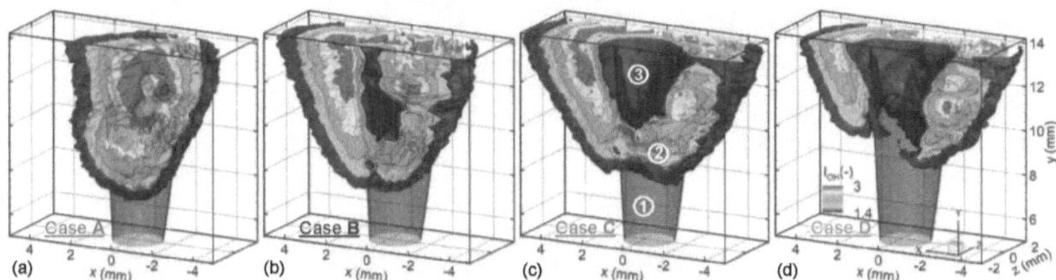

Figure 4.8: Individual visualizations of volatile flames with increasing particle number density denoted as case A, B, C, and D [16].

By defining an intensity threshold referenced to the background, the ignition height H_{ign} was evaluated for different PNDs. Figure 4.9(a) shows a gradual increase of H_{ign} with PND_{ign} calculated in the volume near the ignition position. However, for a reasonable calculation of the ignition delay time t_{ign}, the particle velocity has to be considered. It should be emphasized that the particle velocity can not be assumed to be constant because it significantly changes with PND. The statistical mean velocity profiles were computed conditioned on PNDs. For instance, the mean velocity magnitudes for case A were twice that for case D [16]. Including the PND-dependent velocities, t_{ign} reveal an evident increase with PNDs (see Fig. 4.9(b)).

Furthermore, the topology was divided into three regions: pre-ignition in zone ①, volatile flame in zone ②, and non-flammable region in zone ③. These three zones are schematically shown in Fig. 4.8 for case C. The volume of effective flame region V_2 and the non-flammable region V_3 were evaluated using binary 3D flame structures. The non-flammability ratio R_{nf}

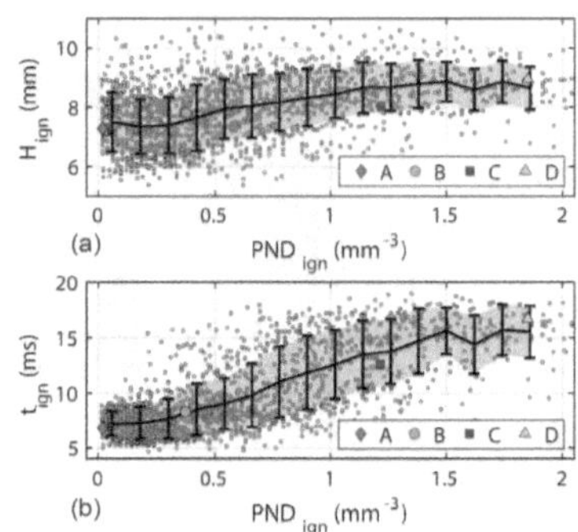

Figure 4.9: (a) Ignition height H_{ign} over PND_{ign}. (b) Ignition delay time t_{ign} over PND_{ign}. Cases A, B, C and D are highlighted [16].

was defined as $R_{\text{nf}} = V_3/(V_2 + V_3)$. Figure. 4.10(a) shows an apparent increase of R_{nf} with $\text{PND}_{\text{flame}}$, in which the solid line indicates a conditional average of R_{nf} values. The flame structures with $R_{\text{nf}} = 0$ have a mean $\text{PND}_{\text{flame}}$ of $0.37\,\text{mm}^{-3}$ (dashed line) representing a flammability-limited seeding density. In Fig. 4.10, a positive correlation of the interaction factor IF_{flame} and $\text{PND}_{\text{flame}}$ is observed. The $\text{PND}_{\text{flame}} = 0.37\,\text{mm}^{-3}$ corresponds to $\text{IF}_{flame} \approx 0.25$, which indicates that the inter-particle distance is approximately $4d_{\text{p}}$. The previous study [15] reports that the flame stand-off distance of single particles is between $1d_{\text{p}}$ and $2d_{\text{p}}$, which means with an inter-particle distance of $4d_{\text{p}}$, the main reaction zones of single-particle volatile flame interfere with each other. The non-flammability can be explained by the intensive particle-particle interaction, which reduces the local gas temperature as well as the oxygen mole fraction.

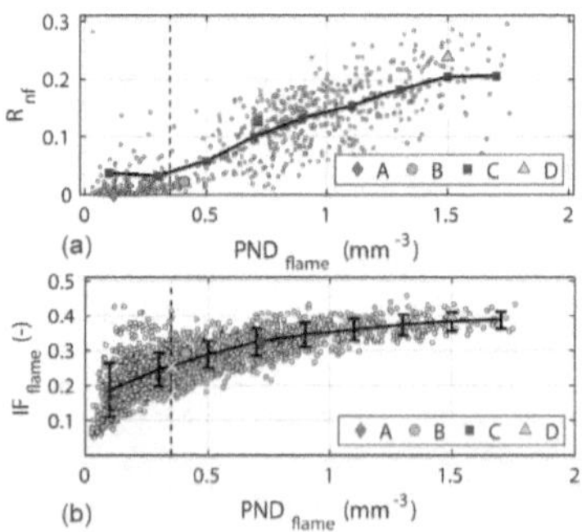

Figure 4.10: Non-flammability ratio R_{nf} over $\text{PND}_{\text{flame}}$. (b) Interaction factor IF_{flame} over $\text{PND}_{\text{flame}}$. Cases A, B, C and D are highlighted [16].

To summarize, the transition effects from single to group combustion of hvb coal particles were comprehensively investigated using volumetric multi-parameter optical diagnostics in **paper VI**. With a statistical analysis conditioned on PND, the particle velocity, ignition delay time, and non-flammability were evaluated, which pointed out the importance of particle-particle and particle-flame interactions in particle group combustion. The critical seeding density or inter-particle distance was determined for the current study, and the flame extinction due to low gas temperatures and rich gas mixtures was discussed. On this basis, other types of fuel particles and different atmospheres will be included in future work to deepen the understanding of particle group combustion of solid fuels.

Chapter 5

Summary and Outlook

5.1 Summary

The present work reported a series of experimental studies firstly for the development of novel volumetric laser diagnostics and secondly on the ignition and volatile combustion of solid fuels using multi-parameter and multi-dimensional laser measurements. Optical imaging techniques were characterized in laminar and turbulent flames and employed to investigate single particle and particle group combustion. Comprehensive data sets, including ignition delay time, volatile combustion duration, flame stand-off distances, particle size, particle number density, and flame topology, were acquired with high accuracy. High-quality results obtained in different air and oxy-fuel conditions allowed for an in-depth understanding of oxy-fuel combustion processes and provided a novel data set for model and simulation validation.

Single-shot tomographic OH-LIF measurements were performed on a premixed CH_4/air flame. It showed that the spatial resolution of tomographic imaging significantly related to the line-of-sight depth. Increasing the thickness of the illumination volume reduced the spatial accuracy of 3D reconstruction. Projection views' number and angle influenced the results as well. As an alternative solution for volumetric imaging, the AOD-based laser scanning was demonstrated using CH_2O-LIF. Quasi-4D measurements successfully visualized lifted flame's structures and allowed for statistical analysis of surface shape characteristics. The laser scanning method was selected for scalar imaging due to its high accuracy and less requirement on equipment, and tomographic imaging was chosen to measure velocity. Both methods were combined to study the flame stabilization of a partially-premixed DME/air lifted jet flame. Simultaneous 3D flame and velocity measurements enabled calculating flame displacement speeds near the stabilization point. Lagrangian analysis of fluid particles allowed characterizing the flame regimes and revealed the essential role of premixing in the stabilization mechanism. With the capabilities demonstrated above, volumetric optical diagnostics were developed for solid fuel combustion studies.

A laminar flow reactor was used to study coal particle combustion experimentally. Its boundary conditions, such as gas temperature and velocity, were extensively character-ized by PIV and OH-LIF measurements. Ignition and volatile combustion of single particle combustion were assessed with simultaneous OH-PLIF, DBI, and LU measurements. The hvb bituminous coal with sizes of $120\,\mu m$ and $205\,\mu m$ were investigated in air (N_2/O_2)

and oxy-fuel (CO_2/O_2) atmospheres. The ignition delay time and volatile combustion duration increased with particle diameters but decreased with oxygen mole fractions. The N_2 replacement by CO_2 resulted in an obvious increase in volatile combustion duration. However, the ignition delay time, especially for large particles, were not significantly impacted by CO_2. Numerical and experimental results revealed that the volatile release could be accelerated with CO_2 due to its high heat capacity and less temperature sink before ignition. It served as compensation for the negative effect of the heat sink for ignition. This work emphasized that gas-phase ignition was a local phenomenon requiring consideration of all relevant parameters and their time evolution. Then, the experiments were extended by employing 3D scanning OH-LIF measurements on single particles. The ignition and volatile combustion were resolved in time and space and scoped on different particle sizes. By evaluating the 3D flame topology, the flame reaction zone's locations were between 1 and 2 particle diameter d_p. Results suggested that larger particles had slower volatile release rates due to lower particle heating rates. Finally, the volumetric OH-LIF measurements were used to study the transition from single particle to particle group combustion. Simultaneous DBI imaging enabled evaluating the particle size, velocity, spatial distribution, and number density. Particles' velocities showed a strong dependence on their number density: velocity decreased strongly in dense streams. Including PND-dependent particle dynamics, the ignition time of particle clouds was clearly delayed by an increasing PND. More interestingly, 3D flame topology revealed non-flammability at higher PNDs. The heavy volatile release resulted in lower gas temperatures and fuel-rich mixtures inside the enveloping volatile flame, which prevented combustion in these zones. The critical PND correlated to an average inter-particle diameter of approximately $4d_p$. Considering the reaction zone's distance of $1{\sim}2d_p$, it implied that individual flames interfered in the case of $4d_p$ distance. With that, flame-flame interaction revealed its importance in flame extinction and particle group combustion.

5.2 Outlook

An essential aspect in particle group combustion study is the in-situ particle location, which remained 2D resolved in previous measurements. To better evaluate the particle-particle interaction, the three-dimensional distribution of particle clouds should be taken into account. Tomographic PTV is a potential approach to address particle trajectories and velocity, which could further combine flame topology visualization in future work. Additionally, particle size and non-spherical shape should be addressed in 3D. It is possible to extend the current highly resolved line-of-sight DBI system by adding more imaging units for a tomographic sizing measurement. The particle rotation before and during the volatile combustion correlates directly to the particle shape and size. Particularly, biomass should be included in future work, which has different volatile content and combustion behavior compared to bituminous coal. To further understand the devolatilization process, the gas temperature and composition in the vicinity of particles should be investigated. For this purpose, back-orientated Schlieren (BOS) is a promising technique based on detecting changes in the refractive index resulting from gas temperature gradients and mixing. The measurement should start from single particles in 2D and gradually extend to particle groups in 3D, and focus on the pre-ignition stage. Furthermore, future experiments should address the soot formation in coal and biomass combustion. The polycyclic aromatic hydrocarbons (PAH) are precursors for soot and could be qualitatively visualized using LIF measurements. Soot particles' volume fractions could be measured by laser-induced incandescence (LII) and calibrated by simultaneous extinction measurements. Combining these techniques with OH-LIF, further understanding of pollutant formation during the volatile combustion could be obtained. Last but not least, fuel blends of coal and biomass should be included in future investigations. Biomass addition will largely change the coal combustion behavior. Different blending ratios and gas atmospheres should be considered for the first investigation in the laminar flow reactor. The methods and results achieved in the present work provide a solid basis for future experimental work and modeling validations.

Appendix

Appendix A (Paper I):
Tomographic OH-LIF measurements in laminar and turbulent flames

Tomographic imaging of OH laser-induced fluorescence in laminar and turbulent jet flames

Tao Li[1], Jhon Pareja[1,2], Frederik Fuest[3], Manuel Schütte[3], Yihui Zhou[4], Andreas Dreizler[1,2], Benjamin Böhm[1]

[1] Institute for Reactive Flows and Diagnostics, Technische Universitäät Darmstadt, Jovanka-Bontschits-Straße 2, 64287 Darmstadt, Germany

[2] Darmstadt Graduate School of Excellence Energy Science and Engineering, Technische Universität Darmstadt, Jovanka-Bontschits-Straße 2, 64287 Darmstadt, Germany

[3] LaVision GmbH, Anna-Vandenhoeck-Ring 19, 37081 Göttingen, Germany

[4] Dalian University of Technology, Linggong Rd. 2, 116023 Dalian, China

E-mail: bboehm@rsm.tu-darmstadt.de

Juli 2017

Abstract. In this paper a new approach for three-dimensional flame structure diagnostics using tomographic laser-induced fluorescence (Tomo-LIF) of the OH radical was evaluated. The approach combined volumetric illumination with a multi-camera detection system of 8 views. Single-shot measurements were performed in a methane/air premixed laminar flame and in a non-premixed turbulent methane jet flame. Three-dimensional OH fluorescence distributions in the flames were reconstructed using the simultaneous multiplicative algebraic reconstruction technique (SMART). The tomographic measurements were compared and validated against results of OH-PLIF in the laminar flame. The effects of the experimental setup of the detection system and the size of the volumetric illumination on the quality of the tomographic reconstructions were evaluated. Results revealed that the Tomo-LIF is suitable for volumetric reconstruction of flame structures with acceptable spatial resolution and uncertainty. It was found that the number of views and their angular orientation have a strong influence on the quality and accuracy of the tomographic reconstruction while the illumination volume thickness influences mainly the spatial resolution.

1. Introduction

In combustion science, laser diagnostics have contributed to understand the complex flame-flow interactions. Traditionally, two-dimensional (2D) techniques such as particle image velocimetry (PIV), Rayleigh scattering, and planar laser-induced-fluorescence (PLIF) are used for measurements of velocity, mixing, and temperature fields as well as for the detection of reaction zones [1]. However, a deeper understanding of

the planar techniques towards volumetric laser measurements able to provide out-of-plane data [2]. For this purpose, two general approaches can be identified based on their laser excitation scheme, namely, multiple laser sheet methods and volumetric laser illumination methods. Both approaches can be implemented in combination with single [3, 4] or multiple cameras [5, 6] for signal detection.

Multiple laser sheets can be generated either by using two or more laser sources [5, 7, 8] or by scanning one single laser sheet through the measurement volume of interest [9–14]. Mechanical mirror scanners (i.e., oscillating and polygonal mirrors) are effective for tracking transient events in combustion and fluid dynamics applications [9, 11–14]. However, their maximal scan frequency is restricted by the inertia of the moving mechanical parts [9]. As an alternative, the laser sheet can be scanned, without moving components, by means of an acousto-optic-deflector (AOD). The capabilities of the AOD approach for quasi-4D laser combustion diagnostics was recently demonstrated on flame front detection and multi-plane PIV measurements [10]. Nevertheless, the maximum repetition rates of commercial lasers limit the AOD approach [10].

In the case of volumetric laser illumination, instead of forming a laser sheet, the beam is expanded and collimated to excite a probe volume. In some applications with limited optical access, a single camera may be desirable for signal detection. Volumetric three-component particle tracking velocimetry (3D/3C-PTV) [3] and holographic PIV [15] are examples in which a single camera is suitable for the technique. However, by increasing the number of cameras for signal detection from different viewing directions, volumetric imaging with high spatial resolution can be achieved. In this approach, the line-of-sight signals, recorded simultaneously by the cameras (i.e., views), are used for a tomographic reconstruction of the measurement volume based on well-known algebraic reconstruction techniques (ART) [16,17]. Different tomographic imaging techniques have been developed from this principle.

Tomographic PIV (TPIV) has been applied to measure 3D/3C velocity fields in non-reactive flows [6], internal combustion engines [7, 18], and turbulent lifted jet flames [19–21] using three to four views. In the case of scalar fields, volumetric flame visualization can be accomplished without laser illumination by means of tomographic chemiluminescence. The applicability of this technique has been demonstrated, for instance, on turbulent jet flames [22–24], opposed jet flames [25] and a matrix burner [26]. Unlike TPIV, for an accurate tomographic reconstruction of scalar fields a high number of views is needed [27], which can be done by rotating a single camera [23] or by coupling the cameras with image doublers [25] or fiber-based endoscopes [28].

Volumetric measurements of scalar fields in combustion diagnostics have been extended to tomographic/volumetric laser-induced fluorescence (Tomo-LIF or VLIF) and laser-induced soot incandescence [29]. Demonstration VLIF experiments have been performed on turbulent jet flows using iodine fluorescence [30] as well as on laminar and turbulent flames using CH radical fluorescence [31]. In those cases, single-shot tomographic reconstructions were performed using a variation of the ART algorithm and compared with traditional PLIF in terms of spatial resolution and accuracy.

Moreover, tomographic imaging in turbulent mixing jets has been demonstrated using acetone Tomo-LIF in single-shot [32] and high-speed [33] measurements. LIF signals from the multiple views were processed applying the multiplicative algebraic reconstruction technique (MART) [17]. The potential of the technique towards quantitative concentration measurements was evaluated in terms of precision, accuracy and spatial and temporal resolutions. Using a very similar approach, the experiments have been extended to high-speed, tomographic laser-induced incandescence measurements of soot in turbulent flames [29]. Recently, Tomo-LIF has been extended to fluorescence measurements of the hydroxyl radical (OH) in a lifted flame [34]. The OH is a well-studied radical which is very useful to mark and track high temperature reaction zones during combustion. Therefore, OH-PLIF has been extensively used to investigate flame structures and dynamics in combustion diagnostics [2, 35, 36].

In this work, a comprehensive evaluation of the feasibility, capabilities, and limitations of Tomo-LIF of OH for the detection of three-dimensional flame structures in combustion diagnostics was performed. In the following sections, a detailed description of the approach, which combines volumetric illumination with a multi-camera detection system and a variation of the MART technique for volumetric reconstruction of the signals, is provided. Using a stable laminar premixed flame, tomographic measurements are compared and validated against results from OH-PLIF. The effect of the configuration of the detection system (i.e., number of views and their angular orientation) as well as the effect of the illumination volume thickness on the reconstruction quality are investigated. Additionally, the spatial resolution and uncertainty of the measurements are evaluated and discussed. Finally, the capabilities of the Tomo-LIF technique are demonstrated in a turbulent non-premixed jet flame.

2. Materials and Methods

2.1. Reference flames

The capabilities of tomographic laser-induced fluorescence were evaluated using a laminar premixed CH_4/air flame. The burner consisted of a tube with an inner diameter of 13 mm and with a sharp-edge tapered nozzle. The length of the tube was 350 mm. Flow rates of methane and air of 0.91 and 8.64 nlpm, respectively, were set to generate a stable laminar flame with equivalence ratio of $\phi = 1$. The bulk velocity of the mixture at the exit of the nozzle was $\sim$1.2 m/s, resulting in a Reynolds number of $\sim$1000.

For demonstration, the tomographic OH-LIF technique was tested in a non-premixed lifted turbulent methane jet flame. The CH_4 jet, emanating from a tube with 8 mm inner diameter and 350 mm vertical length, was surrounded by an air co-flow with 130 mm inner diameter. The fuel nozzle was long enough to provide fully-developed pipe

exit velocities of the jet and co-flow were 10 and 0.2 m/s, respectively. The calculated Reynolds number of the jet was ~5000. Under these operating conditions, the flame had a lift-off height of ~28 mm.

2.2. Tomographic laser-induced fluorescence of OH

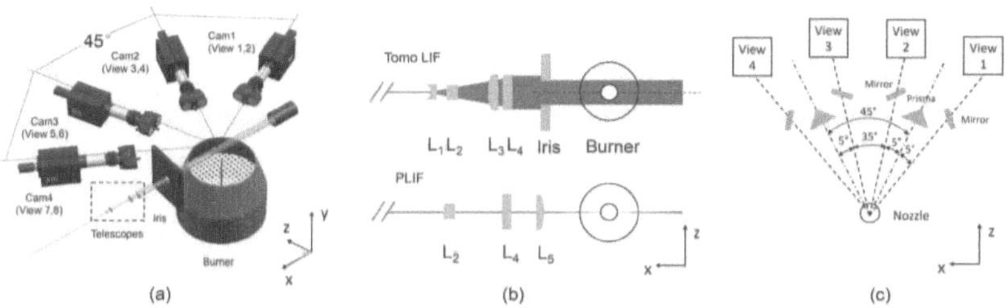

Figure 1. (a) Experimental setup for tomographic LIF measurements. (b) Details of the laser illumination for Tomo-LIF (top) and PLIF (bottom). (c) Schematic of the angular configuration of the image doublers.

The experimental setup for tomographic LIF is schematically illustrated in figure 1(a). A 10 Hz frequency-doubled Nd:YAG laser (Spectra-Physics, PIV400, 350 mJ/pulse) was used to pump a dye laser (Sirah Lasertechnik GmbH, PrecisionScan) operated with Rhodamine 6G. The fundamental wavelength of the dye laser was set to 567.1 nm and frequency doubled with a BBO crystal to excite the $Q_1(8)$ transition of $A^2\Sigma \longleftarrow X^2(\nu' = 1, \nu'' = 0)$ of the hydroxyl radical at $\lambda = 283.55$ nm. After separation of the remaining fundamental radiation using a set of four Pellin-Broca prisms, the UV beam had a pulse energy of 25 mJ at the exit of the laser. As detailed in the upper part of figure 1(b), volumetric illumination was achieved by expanding the laser beam horizontally and vertically with two UV-coated cylindrical telescopes. The first telescope (L_1, $f_1 = -19$ mm and L_3, $f_3 = +200$ mm) expanded the laser beam horizontally (xz plane). After the second telescope (L_2, $f_2 = -25$ mm and L_4, $f_4 = +200$ mm), a collimated, nearly round laser slab with a diameter of ~35 mm was formed. The laser slab was cut by an adjustable sharp-edge iris to form a volumetric illumination with a cross section of 30×30 mm^2 at the measurement volume.

As shown in figure 1(a), the emitted fluorescence signals of OH were simultaneously collected by four identical imaging sets consisting of a CCD camera (LaVision, Imager E-lite), an image intensifier (LaVision, intensified relay optics, IRO), a 100 mm UV lens (Cerco 100, $f/8$, ~5 cm depth of field) and an image doubler (LaVision). The semicircle arrangement, with a radius of ∼450 mm from the measurement volume and a 45° separation angle between each set.

An image doubler, consisting of two high-reflection UV mirrors and one prism, was coupled in front of each UV lens to image two separated views onto each camera. As can be observed in figure 1(c), the angle between the views from each image doubler (e.g., between views 1 and 2) was 10° while the angle between image doublers (e.g., between views 2 and 3) was 35°. The viewing angle, which represents the angle between two outer views (i.e., views 1 and 8), was 155°. Thus, a total of 8 views collected simultaneously fluorescence signal of the same measurement volume but from different perspectives. A high-transmission band-pass filter (LaVision, OH Filter, 320 nm ± 20 nm > 80% transmission) was placed in front of each mirror of the image doublers to suppress background radiation and Rayleigh scattering from the excitation wavelength. The temporal synchronization of cameras, IROs and pump laser was achieved by means of a Programmable Timing Unit (LaVision, PTU X) and IRO controllers (LaVision), controlled with the software DaVis (LaVision, Version 8.3). The intensifiers were gated at 200 ns to suppress the interference from chemiluminescence.

The benchmark measurements were performed in the laminar premixed flame with an effective laser volume of $30 \times 30 \times 30 \, \text{mm}^3$, which was centered to the nozzle and illuminated the whole flame. Thus, the three-dimensional OH distribution in the flame could be fully reconstructed. To evaluate the effect of the size of the illumination volume on the quality of the tomographic reconstruction, the measurements were repeated decreasing the thickness of the illumination volume to 10, 4 and 2 mm while keeping a constant height. This was achieved by operating the laser at maximum output power and symmetrically closing the iris aperture along the z-axis. In this way, the flame was illuminated without variations on the laser energy distribution and energy density.

For comparison and validation, a separated OH-PLIF measurement was performed in the central plane (i.e., xy-plane) of the same laminar flame. As detailed in the bottom part figure 1(b), the laser beam was expanded using the lenses L_2 and L_4 and focused using an additional plano-concave lens (L_5, $f_5 = +300 \, \text{mm}$) into a laser sheet of 30 mm height and 300 μm thickness at the measurement location. The fluorescence signal was collected using the imaging set of camera 3. In this case, the optical resolution of the detection system was determined to be 175 μm by imaging a Siemens-star with 36 segments and calculating the optical modulation transfer function.

2.3. Tomographic reconstruction

The three-dimensional distribution of OH was reconstructed within an effective volume of $30 \times 30 \times 30 \, \text{mm}^3$, located on the laminar flame base to guarantee capturing the fluorescence signal from the entire flame. In the case of the turbulent flame, the volume was located 20 mm downstream from the jet exit, so that the entire flame stabilization region was observed. The image processing was performed using the DaVis software

Prior to the imaging recording, the 8 views were spatially matched using a calibration target (LaVision, Type 7 Target). The data processing included: (1) background subtraction from the raw images, (2) separation of the recorded images into independent views, (3) 2×2 binning, (4) image thresholding, and (5) tomographic reconstruction. The full frame of each camera was 1392×1040 pixels with a resolution of 75 µm/pixel. After symmetrically splitting the images, the size of each view was 696×1040 pixels. To reduce the computational costs, the 8 views were binned by compressing and averaging the intensity of every two neighbor pixels along the x- and y-directions. This resulted in an image size of 348×520 pixels with a resolution of 150 µm/pixel. A 3-5% threshold was introduced to set the background to zero, thus reducing reconstruction artifacts and further decreasing the computational costs.

The tomographic reconstruction was performed using the simultaneous multiplicative algebraic reconstruction technique (SMART) [37] which is implemented in the DaVis software (LaVision, Version 8.3) using parallel computing. For each reconstruction 100 iterations were used. Between individual iterations, a $3 \times 3 \times 3$ voxel average filter with a weighting factor of 0.5 for the surrounding voxels was applied for smoothing. As reference, around eight minutes were needed for the reconstruction of a single-shot, using the 8 full-frame views (i.e., 696×1040 pixels) in a reconstruction volume containing over 150 million voxels of $75 \times 75 \times 75$ µm^3 and 100 iteration steps (with a 12-core computer, 2.6 GHz, 64 GB RAM). The same reconstruction process using the 2×2 binned data, reduced the number of voxels to $\sim$20 million (of 150 µm^3) and decreased the computational cost to around one minute per single-shot.

3. Results and Discussion

3.1. Reconstruction of a laminar premixed flame: reference case

The quality and accuracy of a tomographic algorithm to reconstruct flame structures rely on experimental parameters such as the size of the laser illumination volume and the number and angular orientation of the views. In order to understand the influence of those parameters on the volumetric reconstruction, a reference case using the laminar premixed flame was investigated. In this case, the maximum capacity of the excitation and detection systems was employed (i.e., volumetric illumination of 30mm thickness and 8 detection views).

Figure 2(a) shows a sample sequence of raw images of the OH fluorescence from the laminar flame, captured simultaneously by cameras 1 to 4 (i.e., views 1 to 8). Each 2D image represents the fluorescence intensity integrated along the line-of-sight of the corresponding view. The intensity distribution is a function of the OH concentration, local temperature, Boltzmann fraction, detection system efficiency, beam profile variations, laser absorption, fluorescence trapping, and fluorescence efficiency. Because the aim of the present study is to demonstrate and evaluate the feasibility of the

reaction zones), the intensity maps were not corrected for those effects. Nevertheless, quantitative tomographic OH-LIF measurements would require to take those parameters into account.

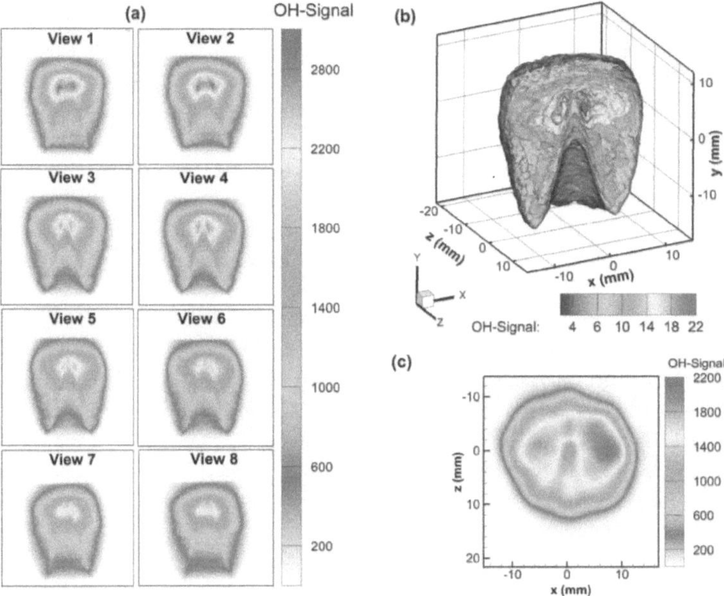

Figure 2. (a) Sample sequence of raw images of the OH fluorescence from the laminar flame, captured simultaneously by cameras 1 to 4. (b) Corresponding single-shot tomographic reconstruction of the signal. For visualization only one half of the volume is shown. (c) Top-view projection of the entire reconstructed signal along the y-axis.

The corresponding reconstructed 3D flame structure, after processing of the 8 views from figure 2(a), is illustrated in figure 2(b). To provide a better view of the internal structure of the flame, only one half of the reconstructed volume is shown. As can be observed, the inner cone of the laminar flame is clearly defined by the steep gradient of the OH signal between the preheat and the reaction zones. The global characteristics of the flame structure are in good agreement with typical single-shot OH-PLIF images of laminar CH_4/air premixed flames stabilized on circular nozzles [39,40]. Figure 2(c) shows the integrated top-view projection, which is the result of the sum of intensities along the y-axis, and provides an overview of OH distribution in space. The characteristic laser absorption in the direction of propagation through the volume (negative direction of the x-axis) can be observed in the image. Additionally, this projection gives an indication

To experimentally validate the tomographic reconstructions, an average of 100 single shots were analyzed and compared against the averaged OH-PLIF results in the laminar flame. Due to experimental limitations, the Tomo-LIF and PLIF measurements were not performed simultaneously. However, the stability and repeatability of the laminar premixed flame provide results which are comparable and useful for validation.

The flame front position, defined here as the location of the steepest OH gradient between the unburned and burned regions, was determined using a Canny edge detection function [41]. For comparison, two volume portions from the Tomo-LIF reconstruction, one parallel and one orthogonal to the plane of the PLIF measurement (xy-plane), were employed. Both volumetric portions had a thickness of 300 μm (comparable to the thickness of the PLIF laser sheet) and were located at the center of the flame. The fluorescence intensity of each portion was averaged along the thickness direction to generate two planar images comparable to the 2D image from OH-PLIF. Averaged flame front profiles were extracted from those three images and are plotted in figure 3(a). As can be seen, the parallel and orthogonal flame front profiles from the Tomo-LIF measurement are in good agreement with the flame front from the PLIF measurement. Additionally, the diameter of the flame base is ∼14 mm, which is in accordance with the diameter of the nozzle.

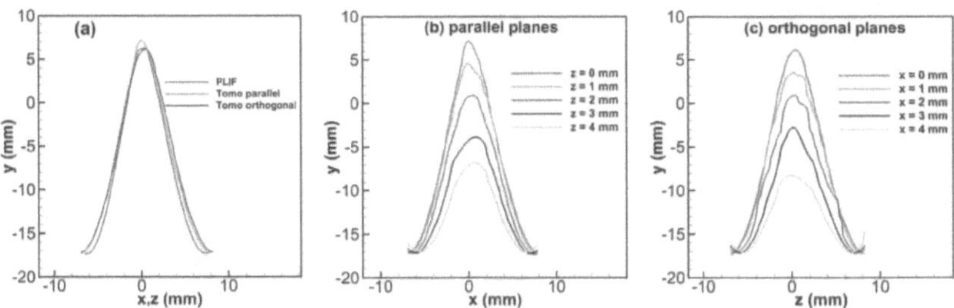

Figure 3. (a) Averaged flame front profiles from 100 single-shots of Tomo-LIF and PLIF at the center of the laminar flame. Averaged flame front contours from Tomo-LIF at different positions along the (b) z-axis and (c) x-axis.

Furthermore, averaged flame front contours were extracted at different planes along the z- and x-axes with a separation of 1 mm and are plotted in figure 3(b) and 3(c), respectively. In this case, a non-linear diffusion filter was applied to smooth the contours. The edge detection was also improved by using a fitting function (smoothing spline method) to overcome the problems related to discontinuous boundaries. The comparison of flame front features from figure 3(b) and 3(c) is an indication of the symmetry expected from a stable laminar premixed flame.

3.2. Tomographic reconstruction: effect of the detection system configuration

The quality of the tomographic reconstruction highly depends on the experimental setup of the detection system. Based on the reference case of section 3.1, the effect of the number of views and their angular orientation on the reconstruction accuracy was investigated. Using an individual single-shot measurement, the tomographic reconstruction was repeated for different configurations of the views, with the same image processing and reconstruction parameters as those of section 2.3. Single-shot 3D reconstructions of the laminar flame using the views 2 to 7, views 3 to 6 and views 1, 3, 6 and 8 are presented in figures 4(a), 4(b) and 4(c), respectively. For visualization, only half of the reconstruction volume is shown in each case. Figures 4(d) to 4(f) show the corresponding top-view projections of the reconstructions.

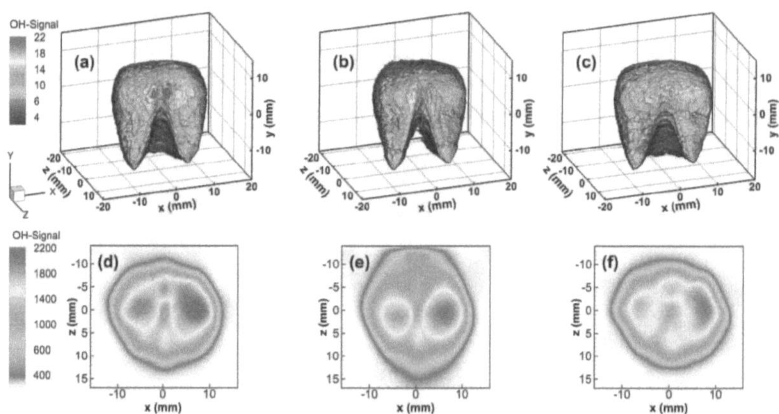

Figure 4. Single-shot reconstructions of the laminar flame using: (a) views 2 to7, (b) views 3 to 6, (c) views 1, 3, 6, 8 along with their corresponding top-view projections (d) to (f). For visualization only one half of the reconstructed volumes are shown in the top row images.

By comparing figures 4(a) and 4(d) with those of the reference case (see figures 2(b) and 2(c)), the SMART algorithm is still able to reconstruct the main structural features of the flame using six views, showing only minor deviations. By processing only the inner views 3 to 6 (i.e., a four-view configuration with a viewing angle of 55°), the flame structure could not be correctly reconstructed. The volumetric reconstruction showed an elongation of the flame cone along the y- and z-axes (see figures 4(b) and 4(e)), suggesting a decrease of the reconstruction accuracy mainly along the thickness direction (i.e., z-axis). However, the reconstruction was improved by rearranging the

4(c) and 4(f) indicate that the flame structure can be resolved with an acceptable quality in all spatial directions when increasing the viewing angle.

3.3. Tomographic reconstruction: effect of the illumination volume thickness

In practice, the size of the laser illumination volume can be restricted by parameters such as maximum laser energy, beam profile and optical access. Therefore, understanding the influence of the laser volume size on the reconstruction quality is essential. For this purpose, tomographic LIF measurements were performed using laser illumination volumes with the same height but with thicknesses of 30, 10, 4 and 2 mm. The laser was still operated at maximum output energy, thus the excitation energy density distribution was similar for all cases. The flame was centered at the middle plane of the laser volume so that the tip of the inner cone was detected. For each case, 100 single-shot volumetric reconstructions, processed as indicated in section 2.3, were averaged and the results are presented in figures 5(a) to 5(d) along with their corresponding top-view projections (figures 5(e) to 5(h)). For visualization, averaged profiles of the flame front were extracted from the central parallel plane (xy-plane) of the reconstruction volumes, as detailed at the end of section 3.1. The results are compared in figure 6.

The reconstructed flames showed high consistency in terms of structure and intensity distribution. With the reduction of the illumination volume thickness, the results of the volumetric reconstruction became smoother while the inner flame cone structure became sharper. This is because the intercepted length of line-of-sight through the laser volume was decreased with the thickness reduction. As consequence, the 2D images of the views contained less information about outer burned gas region, which resulted in an increase of quality of the reconstruction. This is consistent with the high spatial resolution typical of PLIF measurements, which can be interpreted as a tomographic measurement using a very thin illumination volume (thickness in the range of 100 to 500 µm). Nevertheless, the capability of the Tomo-LIF approach to detect three-dimensional flame structures was not highly affected by the size of the laser volume.

3.4. Tomographic reconstruction: spatial resolution and measurement uncertainty

A further analysis of the spatial resolution of the measurements, based on the OH intensity gradients in the reaction zone, is presented in figure 7(a). For all cases of figure 5, averaged intensity profiles were extracted from one branch of the flame along a line in the xy-plane ($z = 0$ mm), at a height of 4 mm above the nozzle exit (dashed lines in figures 5(a) to 5(d)). For clarity, the profiles were horizontally overlapped by setting the location of the points with 50% of the normalized intensity to $x_{50} = 0$ mm, as marked by the black dashed lines in figure 7(a). For validation and comparison, the normalized OH intensity profile of the PLIF measurement is also plotted in figure 7(a).

The results show that the signal intensity gradients increased with the reduction

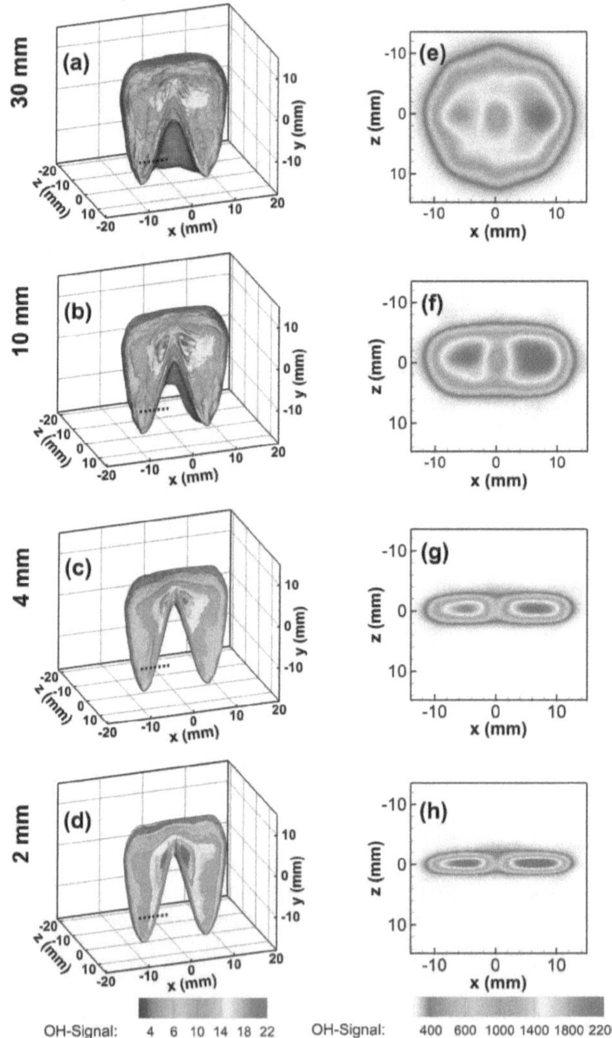

Figure 5. Averaged reconstructions of the laminar flame with an illumination volume thickness of (a) 30 mm (b) 10 mm, (c) 4 mm, and (d) 2 mm and the corresponding top-view projections (e) to (h).

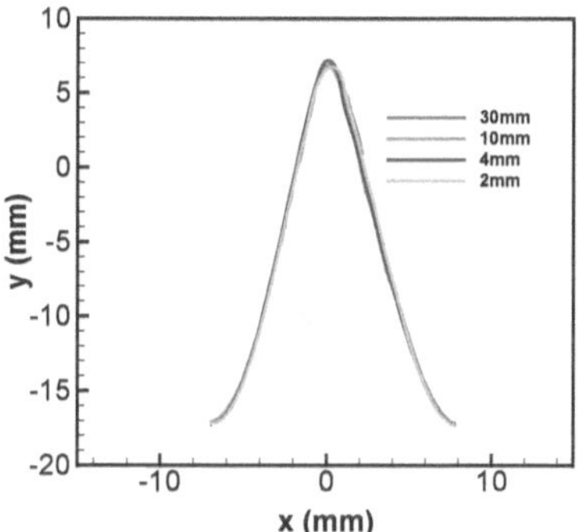

Figure 6. Averaged flame front profiles at the center plane $z = 0$ mm for different laser volume thickness.

Tomo-LIF with 2 mm volume thickness was very similar to the one from PLIF, which means that the spatial resolution decreased with the increase of the illumination volume thickness. In order to estimate the spatial resolution of the Tomo-LIF measurement for the case of the thickest illumination volume (30 mm), the intensity profile from the PLIF measurement was smoothed using a sliding average filter with different filter sizes. As shown in figure 7(b) the PLIF and Tomo-LIF profiles matched at large signal intensities when using a 5×5 Moving-Average filter, and at low signal intensities when using a 7×7 filter. Therefore, the spatial resolution can be estimated as 5 to 7 times lower than the spatial resolution of the PLIF measurement (175 µm). This indicates that, depending on the signal intensity, the spatial resolution of the Tomo-LIF varied between 875 and 1225 µm for an illumination volume of 30 mm thickness.

Figure 8(a) shows a plot of the residual noise distribution in a plane parallel to the xy-plane at $z = 8$ mm, for a single shot reconstruction. To obtain this distribution, the 2D data from the plane at $z = 8$ mm was first smoothed with a 4×4 Gaussian filter and then subtracted from the unfiltered data. The structure of the noise pattern was analyzed by a spatial autocorrelation along the x- and y-directions inside the area of the rectangle (90×90 pixels) marked in figure 8(a). The corresponding autocorrelation functions (ACF) are plotted in figure 8(b). The spatial resolution limit, due to noise, is determined at near-zero AFC, which corresponds to 6 to 8 voxels $= 1050^3$ to 1400^3 µm^3. This range is in good agreement with the spatial resolution calculated from the averaged OH intensity gradients. The histogram of figure 8(c) shows the probability-density-function of the residual noise inside the rectangle marked in figure 8(a). The noise showed a near Gaussian distribution. Using the unfiltered data and the residual noise at $z = 8$ mm, a signal-to-noise ratio (SNR = S/STD) of 28 was determined as the ratio between the averaged signal intensity inside the area marked by the rectangle (S = 7.9) and the

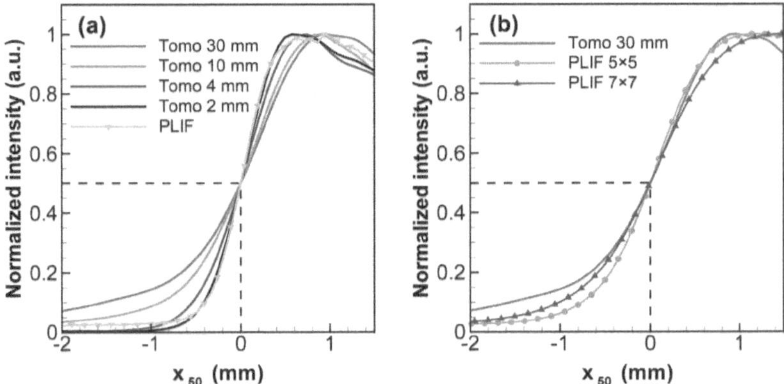

Figure 7. (a) Averaged intensity profiles extracted from one branch of the flame along a line in the *xy*-plane, at a height of 4 mm above nozzle exit (dashed lines in Figures 5(a) to 5(d)). For comparison the averaged profile from the PLIF measurements is included. (b) Filtered intensity profiles from PLIF vs. averaged intensity profile from Tomo-LIF with a volume thickness of 30 mm.

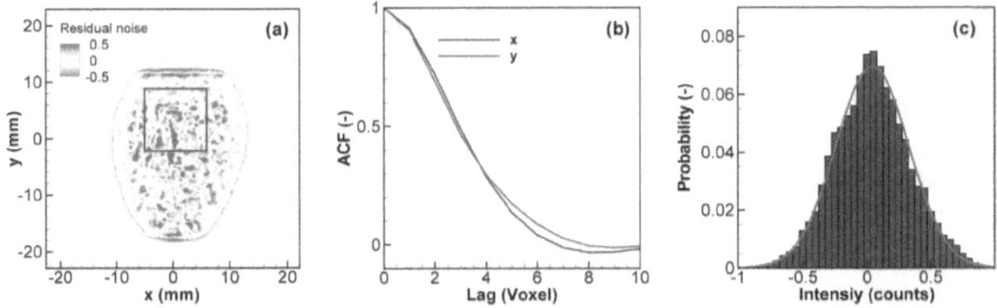

Figure 8. (a) Residual noise distribution in a plane parallel to the *xy*-plane ($z = 8$ mm) from a single-shot Tomo-LIF measurement. (b) Spatial autocorrelation along the *x*- and *y*-directions inside the area of the rectangle marked in (a). (c) Probability-density-function of the residual noise inside the rectangle marked in (a).

3.5. Demonstration of OH Tomo-LIF in a turbulent lifted flame

Figure 10 presents a single-shot tomographic reconstruction of a non-premixed lifted turbulent methane jet flame. For visualization, only a half portion of the 3D flame structure is shown. However, top and side views of the entire flame reconstruction are also provided. The global dimensions, asymmetry and three-dimensional features of the flame base are comparable to those of previous studies in the same burner and with the

central plane of the flame was marked by a circle in figure 9. In this regard, a sequence of 2D images from planes, parallel to the xy-plane, in the reconstructed volume are plotted in figure 10. The images were taken in the vicinity of the marked flame hole at z = -1, 0, 1 and 2 mm to mimic results from PLIF measurements at those locations. The size of that hole structure was estimated to be $3 \sim 4$ mm in diameter, which is resolvable based on the calculated spatial resolution of $\sim$1 mm. By comparing figures 9 and 10, it becomes obvious that 2D flame detection techniques could misinterpret three-dimensional structures as isolated kernels or unconnected flame structures.

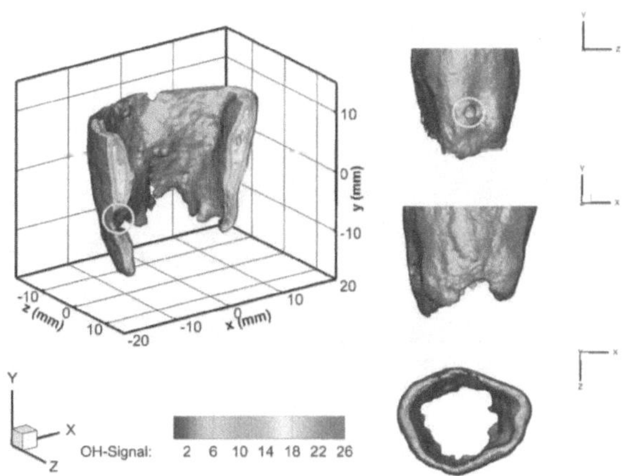

Figure 9. Single-shot tomographic reconstruction of a non-premixed lifted turbulent methane jet flame. For visualization, only a half portion of the 3D flame structure is shown. Top and side views of the entire flame reconstruction are included. The flame feature marked with a circle is discussed in figure 10.

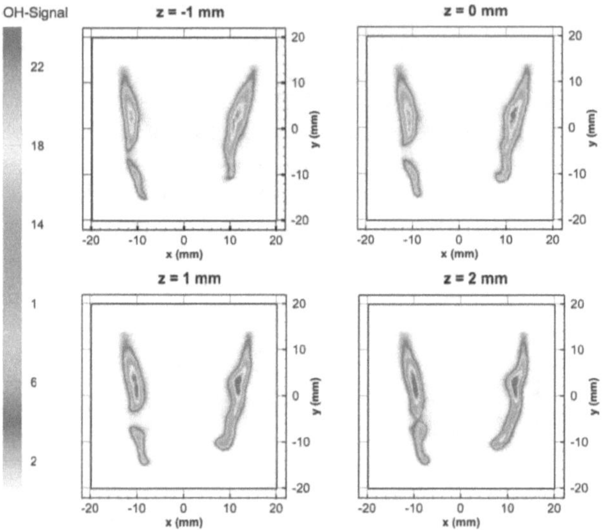

Figure 10. Sequence of 2D images from planes, parallel to the xy-plane, in the reconstructed volume plotted in figure 9. The images were taken in the vicinity of the marked flame hole at $z = $ -1, 0, 1 and 2 mm.

4. Conclusions

Tomographic LIF of OH was demonstrated and evaluated for volumetric reconstruction of 3D flame structures, using volumetric laser illumination in combination with a multi-camera detection system. Single-shot tomographic measurements of a CH_4/air laminar premixed flame were performed with a maximal of eight unique views for detection and a $30 \times 30 \times 30\,mm^3$ laser illumination volume. The comparison and validation of the tomographic measurements against OH-PLIF results proved the feasibility of the OH Tomo-LIF technique for volumetric reconstruction of three dimensional flame features and the volumetric detection of high-temperature reaction zones. After tomographic reconstruction and using two different approaches (comparison against OH-PLIF signals and estimation of the residual spatial noise), a spatial resolution of $\sim$1 mm was consistently calculated. The results showed how the reconstruction quality improved with an increasing number of views and increased viewing angles. A detection setup of four views with an angle of 155° between the outer views was found to yield an acceptable reconstruction quality for the investigated laminar jet flame. Setups with small detection angles were found to cause an elongation of the flame geometry along the line-of-sight direction. Regarding the volumetric illumination, a compromise between

is needed, based on the scales of the flame structures under investigation.

The application of the approach to a non-premixed lifted turbulent jet flame demonstrated the capability to reconstruct complex 3D flame structures and revealed advantages against planar imaging techniques regarding out-of-plane information. Further applications of the technique to more complex configurations, such as confined flames, will require the suppression of laser scattering from surfaces to assure a proper volumetric reconstruction. For the extension to time-resolved Tomo-LIF of OH to investigate transient events in high-turbulence combustion, high-repetition-rate lasers, cameras and intensifiers are required. The low pulse energy of state-of-the-art, high-speed Dye lasers can particularly limit the size of the illumination volume to guarantee high-quality tomographic reconstructions.

Acknowlegements

The authors gratefully acknowledge the financial support by the Excellence Initiative, Darmstadt Graduate School of Excellence Energy Science and Engineering (GSC 1070) and the Deutsche Forschungsgemeinschaft (DFG, German Research Foundation) through SFB-Transregio 129. Andreas Dreizler is grateful for the generous support by the Gottfried Wilhelm Leibniz program of DFG. Yihui Zhou acknowledges the financial support of the China Scholarship Council and Fundamental Research Funds for the Central Universities.

References

[1] Tropea C, Yarin A and Foss J F 2007 *Springer handbook of experimental fluid mechanics* (Springer Science+Business Media, Berlin) 1243

[2] Boxx I, Heeger C, Gordon R, Böhm B, Aigner M, Dreizler A and Meier W 2009 Simultaneous three-component PIV/OH-PLIF measurements of a turbulent lifted, C3H8-Argon jet diffusion flame at 1.5 kHz repetition rate *Proc. Combust. Inst.* **32** 905-12

[3] Peterson K, Regaard B, Heinemann S and Sick V 2012 Single-camera, three-dimensional particle tracking velocimetry *Opt. Express* **20** 9031-37

[4] Fahringer T W, Lynch K P and Thurow B S 2015 Volumetric particle image velocimetry with a single plenoptic camera *Meas. Sci. Technol.* **26** 115201

[5] Shimura M, Ueda T, Choi G-M, Tanahashi M and Miyauchi T 2011 Simultaneous dual-plane CH PLIF, single-plane OH PLIF and dual-plane stereoscopic PIV measurements in methane-air turbulent premixed flames *Proc. Combust. Inst.* **33** 775-82

[6] Elsinga G E, Scarano F, Wieneke B and van Oudheusden B W 2006 Tomographic particle image velocimetry *Exp. Fluids* **41** 933-47

[7] Peterson B, Baum E, Ding C-P, Michaelis D, Dreizler A and Böhm B 2017 Assessment and application of tomographic PIV for the spray-induced flow in an IC engine *Proc. Combust. Inst.* **36** 3467-75

[8] Bode J 2017 Influence of three-dimensional in-cylinder flows on cycle-to-cycle variations in a fired stratified DISI engine measured by time-resolved dual-plane PIV *Proc. Combust. Inst.* **36** 3477-85

[9] Weinkauff J, Greifenstein M, Dreizler A and Böhm B 2015 Time resolved three-dimensional flame base imaging of a lifted jet flame by laser scanning *Meas. Sci. Technol.* **26** 105201

[10] Li T, Pareja J, Becker L, Heddrich W, Dreizler A and Böhm B 2017 Quasi-4D laser diagnostics using an acousto-optic deflector scanning system *Appl. Phys. B* **123** 78-84

[11] Cho K Y, Satija A, Pourpoint T L, Son S F and Lucht R P 2014 High-repetition-rate three-dimensional OH imaging using scanned planar laser-induced fluorescence system for multiphase combustion *Appl. Opt.* **53** 316-26

[12] Brücker C 1995 Digital-Particle-Image-Velocimetry (DPIV) in a scanning light-sheet: 3D starting flow around a short cylinder *Exp. Fluids* **19** 255-263

[13] Böhm B, Kittler C, Nauert A and Dreizler A 2007 Diagnostics at high repetition rates: new insights into transient combustion phenomena textit3rd European Combustion Meeting. Crete, Greece

[14] Wellander R, Richter M and Aldén M 2014 Time-resolved (kHz) 3D imaging of OH PLIF in a flame *Exp. Fluids* **55** 1764-75

[15] Coëtmellec S, Buraga-Lefebvre C, Lebrun D and Özkul C 2001 Application of in-line digital holography to multiple plane velocimetry *Meas. Sci. Technol.* **12** 1392-1397

[16] Herman G T 1980 *Image Reconstruction from Projections - The fundamentals of Computerized Tomography* (Academic Press, New York, USA)

[17] Mishra D, Muralidhar K and Munshi P 1999 A robust MART algorithm for tomographic applications *Numerical Heat Transfer, Part B: Fundamentals* **35** 485-506

[18] Baum E, Peterson B, Surmann C, Michaelis D, Böhm B and Dreizler A 2013 Investigation of the 3D flow field in an IC engine using tomographic PIV *Proc. Combust. Inst.* **34** 2903-10

[19] Osborne J R, Ramji S A, Carter C D, Peltier S, Hammack S, Lee T and Steinberg A M 2016 Simultaneous 10 kHz TPIV, OH PLIF, and CH2O PLIF measurements of turbulent flame structure and dynamics *Exp. Fluids* **57** 65-73

[20] Coriton B, Steinberg A M and Frank J H 2014 High-speed tomographic PIV and OH PLIF measurements in turbulent reactive flows *Exp. Fluids* **55** 1743-62

[21] Weinkauff J, Michaelis D, Dreizler A and Böhm B 2013 Tomographic PIV measurements in a turbulent lifted jet flame *Exp. Fluids* **54** 1624-8

[22] Ishino Y and Ohiwa N 2005 Three-dimensional computerized tomographic reconstruction of instantaneous distribution of chemiluminescence of a Turbulent Premixed Flame *JSME Int. J., Ser. B* **48** 34-40

[23] Worth N A and Dawson J R 2013 Tomographic reconstruction of OH* chemiluminescence in two interacting turbulent flames *Meas. Sci. Technol.* **24** 024013

[24] Weinkauff J, Köser J, Michaelis D, Peterson B, Dreizler A and Böhm B 2014 Volumetric flame measurements in a lifted turbulent jet flame using tomographic reconstruction of chemiluminescence *17th International Symposium on Applications of Laser Techniques to Fluid Mechanics*. Lisbon, Portugal

[25] Floyd J, Geipel P and Kempf A M 2011 Computed Tomography of Chemiluminescence (CTC): Instantaneous 3D measurements and Phantom studies of a turbulent opposed jet flame *Combust. Flame* **158** 376-91

[26] Floyd J and Kempf A M 2011 Computed Tomography of Chemiluminescence (CTC): High resolution and instantaneous 3-D measurements of a Matrix burner *Proc. Combust. Inst.* **33** 751-8

[27] Cai W, Li X, and Ma L 2013 Practical aspects of implementing three-dimensional tomography inversion for volumetric flame imaging *Appl. Opt.* **52** 8106-16

[28] Kang M, Li X and Ma L 2015 Three-dimensional flame measurements using fiber-based endoscopes *Proc. Combust. Inst.* **35** 3821-8

[29] Meyer T R, Halls B R, Jiang N, Slipchenko M N, Roy S and Gord J R 2016 High-speed, three-dimensional tomographic laser-induced incandescence imaging of soot volume fraction in turbulent flames *Opt. Express* **24** 29547-55

[30] Wu Y, Xu W, Lei Q and Ma L 2015 Single-shot volumetric laser induced fluorescence (VLIF) measurements in turbulent flows seeded with iodine *Opt. Express* **23** 33408-18

[31] Ma L, Lei Q, Ikeda J, Xu W, Wu Y and Carter C D 2017 Single-shot 3D flame diagnostic based

[32] Halls B R, Thul D J, Michaelis D, Roy S, Meyer T R and Gord J R 2016 Single-shot, volumetrically illuminated, three-dimensional, tomographic laser-induced-fluorescence imaging in a gaseous free jet *Opt. Express* **24** 10040-9

[33] Halls B R, Gord J R, Meyer T R, Thul D J, Slipchenko M and Roy S 2017 20-kHz-rate three-dimensional tomographic imaging of the concentration field in a turbulent jet *Proc. Combust. Inst.* **36** 4611-8

[34] Halls B R, Hsu P, Legge E, Roy S, Meyer T R, and Gord J R 2017 3D OH LIF measurements in a lifted flame *55th AIAA Aerospace Sciences Meeting, AIAA SciTech Forum*

[35] Böhm B, Heeger C, Boxx I, Meier W and Dreizler A 2009 Time-resolved conditional flow field statistics in extinguishing turbulent opposed jet flames using simultaneous highspeed PIV/OH-PLIF *Proc. Combust. Inst.* **32** 1647-54

[36] Boxx I, Carter C D, Stöhr M and Meier W 2013 Study of the mechanisms for flame stabilization in gas turbine model combustors using kHz laser diagnostics *Exp. Fluids* **54** 1532-48

[37] Atkinson C, Soria J 2009 An efficient simultaneous reconstruction technique for tomographic particle image velocimetry *Exp. Fluids* **47** 553–568

[38] Peterson B, Baum E, Böhm B and Dreizler A 2015 Early flame propagation in a spark-ignition engine measured with quasi 4D-diagnostics *Proc. Combust. Inst.* **35** 3829-37

[39] Hu S, Gao J, Zhou Y, Gong C, Bai X-S, Li Z and Alden M 2017 Numerical and experimental study on laminar methane/air premixed flames at varying pressure *Energy Procedia* **105** 4970-5

[40] Kuhl J, Seeger T, Zigan L, Will S and Leipertz A 2017 On the effect of ionic wind on structure and temperature of laminar premixed flames influenced by electric fields *Combust. Flame* **176** 391-9

[41] Canny J 1986 A computational approach to edge detection *IEEE Transactions on Pattern Analysis and Machine Intelligence* **8** 679-98

Appendix B (Paper II):
Laser scanning CH$_2$O-LIF imaging in a turbulent flame

Experiments in Fluids manuscript No.
(will be inserted by the editor)

High-speed volumetric imaging of formaldehyde in a lifted turbulent jet flame using an acousto-optic deflector

Tao Li · Bo Zhou · Jonathan H. Frank ·
Andreas Dreizler · Benjamin Böhm

Received: date / Revised version: date

Abstract The development of high-speed volumetric laser-induced fluorescence measurements of formaldehyde (CH_2O-LIF) using a pulse-burst laser operated at a repetition rate of 100 kHz is presented. A novel laser scanning system employing an acousto-optic deflector (AOD) enables quasi-4D CH_2O-LIF imaging at a scan frequency of 10 kHz. The diagnostic capability of time-resolved volumetric imaging is demonstrated in a partially premixed DME/air lifted turbulent jet flame near the flame base. Simultaneous imaging of laser beam profiles is performed to account for the laser pulse energy fluctuation and laser sheet inhomogeneity. With the accurate registration of laser sheet positions, the volumetric reconstruction of CH_2O-LIF signals is performed within a detection volume of $17.3 \times 11.9 \times 2.3\,\text{mm}^3$ with an average out-of-plane spatial resolution of 250 μm. A surface detection algorithm with adaptive thresholding is used to determine the global maximum intensity gradient by calculating gradient percentiles. The flame topology characteristics are investigated

T. Li · A. Dreizler · B. Böhm
Institute for Reactive Flows and Diagnostics, Technische Universität Darmstadt
Otto-Berndt-Straße 3, 64287 Darmstadt, Germany

B. Zhou
Department of Mechanics and Aerospace Engineering, Southern University of Science and Technology
No. 1088 Xueyuan Rd., 518055 Shenzhen, China
Combustion Research Facility, Sandia National Laboratories
Livermore, CA 94551, USA

J. Frank
Combustion Research Facility, Sandia National Laboratories
Livermore, CA 94551, USA

T. Li E-mail: tao.li@rsm.tu-darmstadt.de ·
B. Zhou E-mail: zhoub3@sustech.edu.cn ·
J. Frank E-mail: jhfrank@sandia.gov ·
B. Böhm E-mail: boehm@rsm.tu-darmstadt.de ·
A. Dreizler E-mail: dreizler@rsm.tu-darmstadt.de

by evaluating the 3D curvatures of CH_2O surfaces. Curvatures calculated using 2D data systematically underestimate the full 3D curvature due to the lack of out-of-plane information. The inner surfaces near the turbulent fuel jet exhibit higher probabilities of large mean curvature than the outer surfaces. The saddle and cylindrical structures are dominant on both the inner and outer surfaces, and the elliptic structures occur with lower probability. The results suggest that the damping of turbulent fluctuations by the temperature increase through the CH_2O region reduces the curvature, but the local structure topology remains self-similar.

Graphic abstract

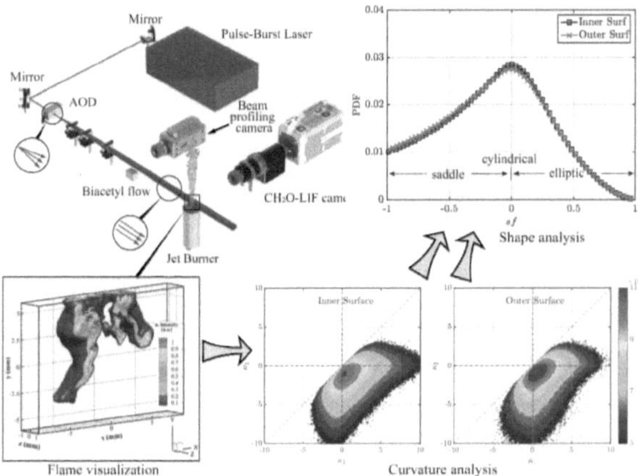

1 Introduction

Laser-based techniques play a vital role in non-intrusive scalar and velocity measurements for investigating flame-turbulence interactions. Diagnostic techniques such as particle image velocimetry (PIV) and laser-induced fluorescence (LIF) have been employed extensively for two-dimensional measurements. However, turbulent flows are inherently three dimensional and evolve in time. Conventional two-dimensional imaging measurements lack information on the out-of-plane motion and the temporal evolution of turbulent flows, leading to ambiguities in data interpretation (Boxx et al., 2009). In order to understand the inherent multi-dimensional transient phenomena in turbulent reacting flows, extension of planar single-shot measurements to high-speed volumetric measurements is required.

Approaches to accommodate laser-based volumetric measurements have been continuously developed in the recent years, and can be classified into two categories, namely volumetric laser illumination and the multiple laser sheets approach. For the volumetric laser illumination, the laser beam is shaped into a collimated laser slab of a few millimeters in thickness and multiple cameras from various viewing angles are employed for simultaneous signal detection.Tomographic reconstructions from the collected line-of-sight signals are performed using various algorithms, such as algebraic reconstruction techniques (ART) (Ma et al., 2017a,b; Wu et al., 2015) and simultaneous multiplicative algebraic reconstruction techniques (SMART) (Halls et al., 2017a; Li et al., 2018; Pareja et al., 2019). For example, high speed volumetric velocity measurements using tomographic PIV have been successfully demonstrated in various turbulent flames (Coriton et al., 2014; Zhou and Frank, 2019) and in-cylinder flows of IC engines (Baum et al., 2013; Peterson et al., 2017). Single-shot and high-speed volumetric scalar measurements of a number of key species such as CH (Wu et al., 2015), OH (Halls et al., 2017b; Li et al., 2018; Pareja et al., 2019) and fuel tracers (Halls et al., 2016, 2017a; Ma et al., 2017b) have been realized in both laminar and turbulent flames. To obtain a sufficient signal-to-noise ratio (SNR) as well as reasonable spatial and temporal resolutions, volumetric measurements are constrained by the availability of high laser repetition rates, high pulse energy, beam profile homogeneity and the number of camera views. Obtaining high laser repetition rates and high pulse energy simultaneously can be a challenge not only from the aspect of laser manufacturing but also for optical components to sustain tremendous amounts of laser power. In addition, the obtainable spatial resolution is typically limited by the number and the orientation of camera views (Li et al., 2018). Despite the fact that the tomographic reconstruction process involves solving an inherently underdetermined problem, 6 - 8 camera views are typically required to obtain a spatial resolution smaller than 1 mm in laminar flames with a probe volume of $30 \times 30 \times 30\,\text{mm}^3$ (Li et al., 2018). Practical obtainable spatial resolutions in turbulent flames are expected to degrade further as the highly convoluted turbulent flame structures add another level of complexity for tomographic reconstruction. The above-mentioned limitations substantially constrains the implementation of such volumetric laser illumination approaches in reacting flows.

For the multiple laser sheet approach, the probe volume is sliced by multiple parallel laser sheets, and a time interval that is smaller than the characteristic flow time scale is used to separate the laser sheets in time. A single camera that is synchronized with the multiple laser sheets is typically employed to record the images from various planes that are used for volumetric reconstruction. Multiple laser sheets can be achieved by using multiple static laser sources (Bode et al., 2017; Peterson et al., 2015; Shimura et al., 2011; Trunk et al., 2013), multiple wavelengths generated from a single source (Frank et al., 1991) or a single laser sheet rapidly sweeping through the volume of interest. The latter approach, which is referred to as the laser scanning technique, has been realized in the past by using mechanical mirrors, i.e. oscillating mirrors (Kychakoff et al., 1987; Yip et al., 1988) and rotating polygonal mirrors (Patrie, 1994). Among recent investigations, kHz-rate imaging has been used for flow field measurements (Wellander et al., 2011), mixture fraction visualization (Miller et al., 2014), and flame front and reaction zone detection (Olofsson et al., 2006; Weinkauff et al., 2015; Wellander et al., 2014). Compared to the volumetric laser illumination approach, the multiple laser sheet approach offers a better in-plane spatial resolution, less constraints on the laser pulse energy and the number of camera views. However, in addition to the laser safety issue (Weinkauff et al., 2015), the intrinsic limitations associated with the mirror-based scanners are the high mass inertia and instability of the moving mechanical parts, which restricts the accuracy and precision of the laser beam position, especially at high scanning speeds.

A recently demonstrated alternative to mirror-based scanners is multiple laser sheet generation using an acousto-optic deflector (AOD), which does not contain any moving parts and is therefore advantageous over the mirror-based scanners in terms of scan frequency, accuracy, precision and spatial resolution (Römer and Bechtold, 2014). Employing 532 nm laser radiation, kHz-rate volumetric visualizations using an AOD have been demonstrated in a turbulent lifted jet flame by means of Mie-scattering (Li et al., 2017) and in-cylinder flow measurements in an IC engine via multi-plane PIV (Bode et al., 2019). However, 4D mapping of chemical species such as OH radicals and CH_2O in turbulent flames using an AOD scanning system has not been reported in the literature. The LIF excitation of such intermediate species often requires an excitation wavelength in the UV range which imposes additional requirements on the AOD crystal material and restrictions on the maximal obtainable scan angle. In addition, to capture motions of higher Reynolds-numbe flow, laser repetition rates that are significantly higher than 10 kHz are typically desired.

The present work demonstrates the recent advancement in quasi-4D laser-induced fluorescence measurements of formaldehyde (CH_2O-LIF) in a partially premixed DME/air lifted turbulent jet flame by employing an AOD scanning system combined with a pulse-burst laser operated at a repetition rate of 100 kHz. In the following sections, the experimental methods are first introduced. The CH_2O reconstruction is performed and the three-dimensional CH_2O surface detection with an adaptive thresholding method is then described. An evaluation of flame topology and curvatures from the 3D CH_2O-LIF measurements is performed to characterize the flame structures close to the base of the lifted turbulent flame.

2 Experimental Methodology

2.1 Acousto-optic deflection

As schematically illustrated in Fig. 1, a typical AOD scanning system consists of a function generator, a radio frequency (RF) driver and an acousto-optic crystal. The piezo-transducer attached on the crystal surface is driven by RF-signals to generate a high frequency acoustic wave propagating through the crystal. The acoustic frequency f is directly controlled by the DC voltage assigned by the function generator. The acoustic wave induces local rarefaction and compression in the crystal, leading to periodical changes in the material density. As a consequence, the local refractive index varies periodically and the crystal serves as an optical diffraction grating. When the AOD is operated in the Bragg regime, the interaction length L that the laser travels through the crystal must satisfy

$$L \gg \frac{\Lambda^2}{\lambda} \tag{1}$$

where Λ denotes the acoustic wavelength and λ is the wavelength of the incident laser. The maximum intensity of the first diffraction order is then located at a particular incident angle $\theta_i = \theta_B$, where θ_B is the Bragg angle that is a function of the acoustic frequency (f) and acoustic velocity (v) in the crystal as expressed in equation (2),

$$\theta_B \approx sin\theta_B = \frac{\lambda f}{2v} \tag{2}$$

The separation angle θ_d is twice the Bragg angle, namely

$$\theta_d \approx \frac{\lambda f}{v} \tag{3}$$

In the present measurements, λ and v are constant and θ_d is solely determined by the acoustic frequency f, which shows nearly linear response to the DC voltage level applied. The AOD used in the current work comprised a water-cooled quartz crystal (D1340, ISOMET) and a tunable RF-driver (RFA333, ISOMET). The DC voltage in a range of 0 - 10 V was generated from an arbitrary waveform generator (33500B Series, Keysight), resulting in a frequency bandwidth, Δf, of 40 MHz and a maximum scan angle $\theta_s \approx \lambda \Delta f / v$ of 2.3 mrad. A maximum diffraction efficiency of 70% in first diffraction order was obtained and optimized for 355 nm at the center acoustic frequency f_c= 120 MHz. For other acoustic frequencies, the diffraction efficiency remained above 60% over the complete range of scan angles.

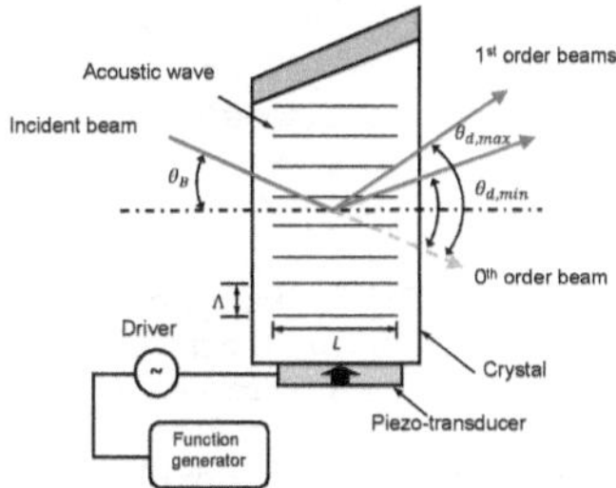

Fig. 1: Schematic of the acousto-optic deflector system with the laser deflection in the Bragg regime (Bragg angle θ_B). First order beam shows a separation angle range between $\theta_{d,min}$ and $\theta_{d,max}$ as the acoustic frequency varies from f_{min} and f_{max}. Scan angle $\theta_s = \theta_{d,max}$ - $\theta_{d,min}$.

2.2 Laser-induced fluorescence of CH_2O

To demonstrate the feasibility and capability of a 100 kHz pulse-burst laser combined with an AOD scanning system for quasi-4D CH_2O-LIF imaging, a partially premixed lifted turbulent DME/air jet flame was investigated. CH_2O is a key intermediate species that marks the fuel decomposition/consumption process (Gabet et al., 2013). The entire experimental setup is schematically shown in Fig. 2. The burner consisted of an air co-flow of 76 mm inner diameter and a central nozzle of 2.5 mm inner diameter supplied with a fuel rich DME/air mixture. The co-flow bulk velocity was 0.2 m/s, and the bulk velocity at the exit of the nozzle was approximately 17 m/s, resulting in a Reynolds number of approximately 4500. The jet DME/air mixture had an equivalence ratio of 9.5, forming a lifted flame with an average lift-off height of approximately 25 mm above the jet exit. The flow field was measured by particle image velocimetry (not shown here), and the mean gas velocity at the flame stabilization position was 0.5 m/s. The flame propagation during a scan sequence (100 μm) was estimated to be 0.05 mm.

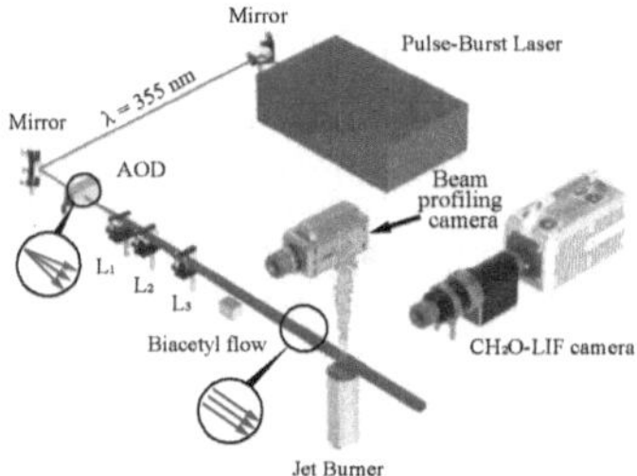

Fig. 2: Experimental setup for quasi-4D CH_2O-LIF measurements in a turbulent lifted DME/air jet flame using an AOD scanning system.

The CH_2O-LIF was excited using the third harmonic ($\lambda = 355$ nm) of the pulse-burst laser, which provided a 5 ms duration burst of pulses at 100 kHz. The laser pulses were synchronized with the AOD scanning system by a function generator that supplied a voltage step function (Li et al., 2018) with a 10 μs step width and a 100 μs cycle period to the AOD. This provided a 10 kHz scanning frequency with 10 planes in each scan cycle. The average pulse energy was 2.0 mJ, and the pulse duration was 7 ns. A UV-coated cylindrical telescope composed of a plano-concave lens (L_1, $f_1 = -350$ mm) and a plano-convex lens (L_2, $f_2 = +750$ mm) expanded the laser beam vertically. With a plano-convex cylindrical lens (L_3, $f_3 = +1000$ mm), the laser sheets were parallelized and focused to a thickness of 100 μm. The CH_2O fluorescence signal was collected perpendicularly to the laser sheets by a high-speed CMOS camera (Fastcam SA-X2, Photron) coupled with a high-speed lens-coupled image intensifier (HICATT, Lambert Instruments). The intensified camera was equipped with a 58 mm lens (Nikkor, $f/2.8$, depth of field > 3 mm). A 500 nm (Schott) short-pass filter and a 355 nm notch filter (Semrock) suppressed flame luminosity and laser scattered light. The CMOS camera was operated at 100 kHz with 384×264 active pixels and was synchronized with the image intensifier, the pulse-burst laser and the AOD scanning system. Individual CH$_2$O-LIF images were recorded with 75% of the maximum intensifier gain and a 250 ns gate to suppress background luminosity. Using these settings, signal cross-talk from flame luminosity was negligible as compared to the CH$_2$O fluorescence signal. The recorded images exhibited an in-plane projected pixel size of 45 µm and a field of view (FOV) of 17.3×11.9 mm^2. The variation in FOV size was negligible and was examined by traversing an imaging target along the z-axis. The imaging resolution was evaluated to be approximately 175 µm using USAF resolution test chart.

2.3 Beam profile imaging and laser position registration

To account for energy fluctuations as well as laser sheet inhomogeneities for individual pulses from a burst, the beam profile was simultaneously imaged using an additional beam profiling CMOS camera (Phantom 7.3) with an integrated fiber-coupled intensifier (P24, Hamamatsu). The beam profiling camera was equipped with a 58 mm lens (Nikkor, $f/2.8$, depth of field > 3 mm) and a 355 nm notch filer. Beam profiles were recorded by LIF imaging in a laminar flow of air and biacetyl (C$_4$H$_6$O$_2$) vapor that was excited by the same 355 nm laser sheets as shown in Fig. 2. The biacetyl flow was placed in front of the jet burner, and the loss in laser pulse energy via biacetyl vapor absorption was measured to be negligible ($< 2\%$). Operating at 100 kHz with 500 ns gate, the beam profiling camera provided a frame size of 64×256 pixels and a pixel resolution of 56 µm/pixel. A snapshot of biacetyl fluorescence is shown in Fig. 3(a), from which the laser sheet profile plotted in Fig. 3(b) was obtained by horizontally averaging the recorded image.

The accurate position registration of the individual laser sheets was essential for accurate three-dimensional reconstruction. The parallelization of the multiple laser sheets and their thickness in the focus plane were measured on a beam monitor camera (WinCamD-LCM, Dataray). Figure 3(c) shows relative positions of the 10 parallel tightly focused laser sheets at the focus plane and the corresponding vertically averaged intensity profile is shown in Fig. 3(d). The total scan depth was measured to be 2.3 mm. To examine the parallelization of the laser sheets, the beam monitor camera was traversed approximately ±15 mm along the laser propagation direction (x-axis) across the measurement volume. The maximum deviation of the total scan depth was less than 100 µm ($\sim4\%$ of scan depth), indicating a reliable parallelization between laser sheets. The impact on reconstruction accuracy was considered to be negligible.

The discretization of illumination planes along the z-axis was not equidistant since the separation of the planes depended on the incident angle of the laser sheets to the converging lens (L_3). The z-discretization between the laser sheets varied from 198 µm to 286 µm (250 µm on average) depending on the laser sheet position. The size of the largest vortical structures in the flow field were estimated by the integral length scale at the flame base using $L_{base} \approx 0.07 \times l_{lift-off} = 1.75$ mm (Gautam, 1984). The average out-of-plane spatial resolution was approximately 1/7 of the integral length scale and thus, was sufficient to resolve these structures.

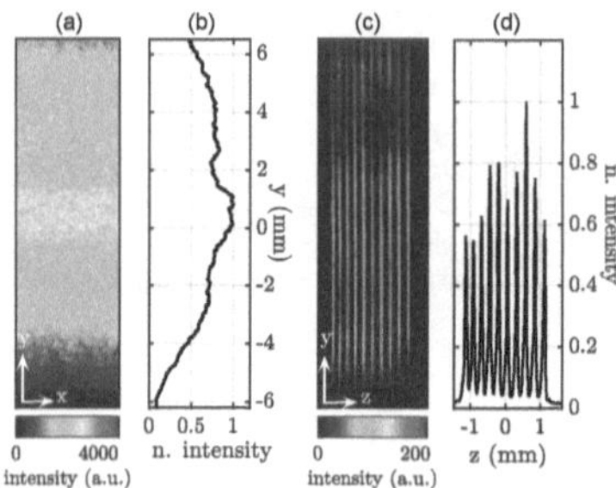

Fig. 3: (a) Single-shot biacetyl fluorescence image and (b) averaged vertical intensity profile. (c) Recorded positions of the parallel laser sheets at the focus planeand and (d) average horizontal intensity profile.

3 Data Processing Procedures

3.1 CH$_2$O-LIF signal reconstruction

For the signal reconstruction, the raw CH$_2$O-LIF images were first corrected for shot-to-shot laser energy fluctuation and sheet intensity inhomogeneity. The images were then filtered with 3×3 median and 3×3 Gaussian filters to reduce noise. To estimate the SNR, a laminar premixed DME/air flame that yielded comparable CH$_2$O-LIF signals to the turbulent jet flame was imaged with the same detection system. A SNR of approximately 15 was evaluated in a homogeneous CH$_2$O region without applying image filters. With the accurate position of each individual laser sheet as registered by the beam-position camera, the raw images from each of the ten planes were assigned to the corresponding z-positions to form a 3D signal matrix representing an entire detection volume of $17.3 \times 11.9 \times 2.3\,\mu m^3$. The volumetric matrix was linearly interpolated in the z-direction such that the spacing between the adjacent z planes was approximately $45\,\mu m$ which was identical to the spatial pixel resolution in the x-y plane. The entire probe volume was then represented by $384 \times 264 \times 50$ voxels of $45\,\mu m^3$ and the CH$_2$O-LIF intensity was saved in a volumetric matrix V_{raw}.

Figure 4 shows a single-shot 3D structure represented by the iso-contours of CH$_2$O-LIF signal intensity. The coordinate system (x, y, z) is centered in the reconstruction volume, which is approximately 25 mm downstream of the jet nozzle exit. Two branches of the highly convoluted CH$_2$O structures can be readily observed.

As the fuel-rich jet mixes with the ambient air up-stream, mixtures approach stoichiometric conditions near the stabilization point which can be approximated as the lowest point in each branch of the 3D CH$_2$O structures (Gordon et al., 2012). Near the stabilization point, in the regions towards the fuel jet stream the mixtures remain fuel-rich, while mixtures towards the ambient flow are diluted by mixing with ambient air resulting in a fuel-lean mixture (Lawn, 2009; Lyons, 2007). However, the typical structure of a triple flame (Peters, 2000) was not observed in this measurement. The two premixed branches are probably collapsed on the non-premixed tail (Karami et al., 2016; Kioni et al., 1993) and the CH$_2$O structure resembles an edge-flame (Buckmaster, 2002).

It can be seen from Fig. 4 that the inner layer (surfaces nearest to the jet axis) of the CH$_2$O structures is more wrinkled than the outer layer (surfaces nearest to the co-flow). The flame surface wrinkling and curvature are indicators that show the extent to which turbulence-flame interactions impact the flame topology, and the following data analysis focuses on the flame surface shape characteristics. The first step in this analysis is to detect the locations of the 3D flame surfaces.

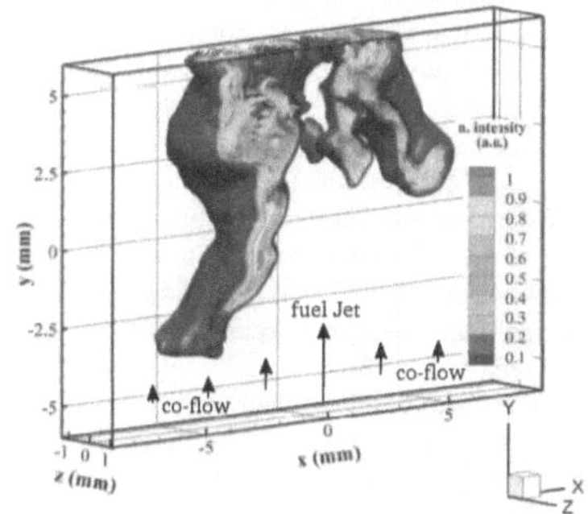

Fig. 4: Single-shot CH$_2$O-LIF signal reconstruction

3.2 CH$_2$O surface detection using adaptive thresholding

For 2D scalar measurements, edge detection is typically accomplished either by employing gradient-based algorithms such as the Canny method (Canny, 1986) or by identifying an iso-intensity contour with fixed intensity thresholding. However, the local maximum of the three-dimensional scalar gradients can be highly dispersed in space, and a simple correlation between an intensity threshold and local scalar gradient maximum cannot be easily determined (Pareja et al., 2019). To properly determine the boundary of a 3D CH$_2$O structure, we developed a surface-detection algorithm using a similar approach to that proposed in the previous work (Pareja et al., 2019). The location of the CH_2O surface was defined here as a 3D iso-intensity contour at the intensity level (I_{cont}) that coincides with the global gradient percentile as described below.

First, the 3D CH_2O-LIF signal matrices V_{raw} were smoothed with a $3 \times 3 \times 3$ Gaussian filter to reduce noise, and the 3D CH_2O-LIF intensity gradients were calculated using a second-order central difference. For each instance, the 3D CH_2O-LIF intensity gradients were stored in an array and sorted by magnitude in ascending order in which zero gradient was not considered. A percentile gradient was determined from the value that is ranked as the ith percentile of the array. To select the percentile used to determine the intensity threshold, a sensitivity analysis was conducted as follows. Figure 5(a) shows an example of a plane of the 3D CH_2O structures with iso-gradient contours that correspond to the 10th (P_{10}), 20th (P_{20}), 30th (P_{30}), 40th (P_{40}) and 90th (P_{90}) percentile gradient values. The mean intensities I_{cont} were then calculated as the average LIF signal within the region enclosed by the corresponding iso-gradient contours.

In Fig. 5(b), the joint probability density function ($JPDF$) of normalized CH_2O-LIF intensity and its gradient from a 3D CH_2O matrix is plotted. A polynomial is fit to the non-zero $JPDF$ data points as shown by the solid line. In addition, the intensities I_{cont} (vertical dotted lines) and their projected points on the fitted curve are marked (solid circles). The peak location of the fitted curve is marked with a yellow cross, which is the representative global gradient maximum of this matrix. The intensity obtained from P_{20} correlates well with the global gradient maximum. Using various I_{cont} for thresholding, edge detection is performed with selected intensity iso-lines to test sensitivities. The intensity thresholds of $I_{cont}(P_{90})$ leads to the largest uncertainties in terms of detection of the CH_2O boundary location. This observation is consistent with the $JPDF$ distribution, in which high intensities are correlated with low gradients. Near $I_{cont}(P_{20})$ and $I_{cont}(P_{30})$, the boundary shows little sensitivity to the percentile value chosen. This is because these intensities correlate well with the global gradient maximum, which is in accordance with the $JPDF$ analysis. Fig. 5(c) shows additionally that the boundary determined by P_{20} within the probe volume appear as connected, and flame structures of various scales are well represented. As a result of this sensitivity study, the P_{20} percentile is chosen as the threshold value for surface detection in this work. Since no fitting is required, the percentile calculation is much faster than direct computing of global gradient maximum from $JPDF$s but still provides a reasonable approximation.

Figure 6(a) shows a single-shot 3D surface which corresponds to the CH_2O structure in Fig. 4. Because the gradient value of P_{20} percentile is estimated for each single-shot CH_2O-LIF matrix, the intensity threshold for edge detection is automatically adapted to the signal level of each CH_2O structure. Due to the gradient-based character, this algorithm provides a reliable detection of the CH_2O surface that is not sensitive to intensity fluctuations and background noise.

The detected flame surface is shown along with the raw CH_2O-LIF signal in Fig. 6(b) for a 2D slice that is extracted from the 3D reconstruction. Due to the fact that the surface wrinkling on both sides of the CH_2O region are different, the CH_2O boundary was separated into an inner and outer surface. Figure 6(b) shows inserted break points

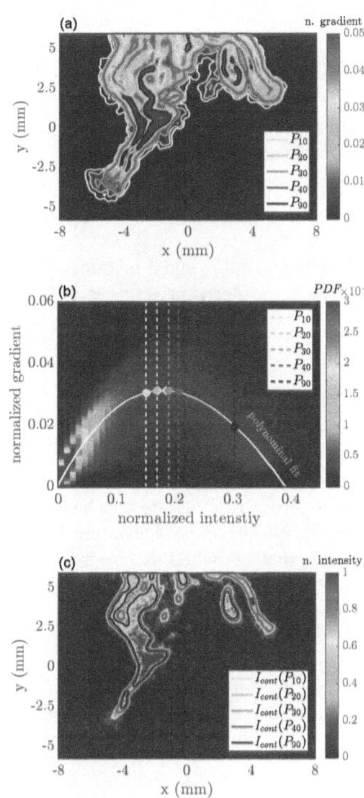

Fig. 5: Cross-section of a single-shot CH$_2$O-LIF signal reconstruction shown as (a) gradient map with gradient iso-lines of percentiles P. (b) *JPDF* of intensity and gradient. (c) The corresponding intensity map with intensity iso-lines I_{cont} corresponding to gradient percentiles.

that correspond to the lowest axial position on each side of the flame base. In the following, separate structural analysis is performed on the inner and outer surfaces.

3.3 Local surface topology

The local surface topology is investigated by evaluating the CH$_2$O-LIF iso-surfaces that are detected using the adaptive thresholding method discussed in section 3.2. The vector n normal to an iso-surface is defined as

$$n = \frac{\nabla I_{cont}}{|I_{cont}|} \tag{4}$$

, where the normal vector points towards regions of higher intensity. By computing the three invariants (I_1, I_2 and I_3) of the curvature tensor, the principal curvatures, κ_1 and κ_2 can be obtained as the two eigenvalues of the characteristic equation (Han et al., 2019)

$$\kappa^2 + I_1\kappa + I_2 = 0 \tag{5}$$

The principal curvatures, κ_1 and κ_2, are the numerically maximum and minimum curvatures, such that $\kappa_1 > \kappa_2$. The Gaussian curvature κ_g and the mean curvature, κ_m, can be computed as

$$\kappa_g = \kappa_1\kappa_2 \tag{6}$$

$$\kappa_m = \frac{\kappa_1 + \kappa_2}{2} \tag{7}$$

As schematically shown in Fig. 7, the local surface geometries of CH_2O surfaces can be described by evaluating the $\kappa_m - \kappa_g$ *JPDF*. The structure topology can be classified as elliptic ($\kappa_g > 0$), cylindrical ($\kappa_g = 0$) or saddle ($\kappa_g < 0$) shape. According to the positive and negative value of κ_m, the shape is further classified as convex ($\kappa_m > 0$) or concave ($\kappa_m < 0$) toward regions of higher CH_2O-LIF signal. The region $\kappa_g > \kappa_m^2$ implies non-physical complex curvatures. To minimize the artifacts caused by the pixel discretization, the raw volumetric CH_2O scalar fields V_{raw} were further processed with local Gaussian-smoothing with filter size of 5 voxels. To eliminate the boundary effect on the curvature statistics, points near the edge of the detection volume were excluded from this analysis.

4 Turbulent Flame Topology

In this section, 3D spatially resolved data are used to investigate the topography of the lower part of the lifted jet flame including the flame base. Curvatures calculated using 2D and 3D data are compared and flame structures are characterized by surface curvature statistics. Finally, a time-resolved CH_2O 3D reconstruction is shown to demonstrate the temporal evolution of the lifted flame.

4.1 CH_2O surface curvatures

The three-dimensional CH_2O structures are investigated by evaluating curvatures of CH_2O iso-surfaces, as described in section 3.3. Although this method is employable for 2D and 3D surfaces, in 2D case κ_1 and κ_2 are equal to κ_m. Since the 2D curvature has been widely used to examine flame wrinkling in previous studies, it is of interest to compare the mean curvature as obtained from 2D and 3D data. In Fig. 8, the probability density function (*PDF*) of the two-dimensionally computed κ_m is compared

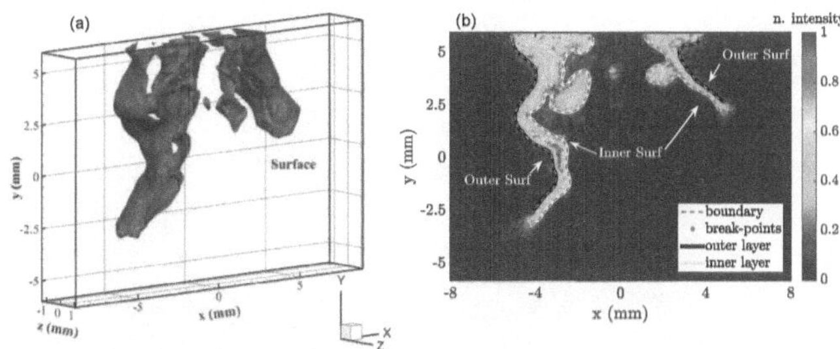

Fig. 6: (a) Single-shot surface detection of CH$_2$O structure in Fig. 4, which is evaluated with adaptive boundary detection based on a gradient percentile P_{20}. (b) 2D slice of single-shot CH$_2$O-LIF signal with the determined boundaries. The boundary is separated into the inner and outer layer surface by inserting break-points (red dots).

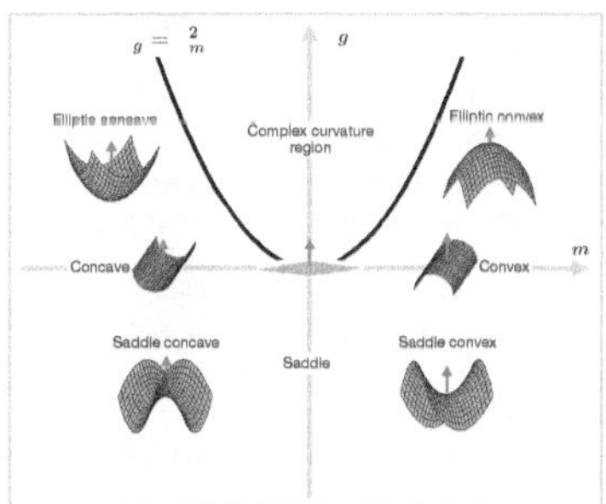

Fig. 7: Classification of local CH$_2$O surface shapes within the $\kappa_m - \kappa_g$ diagram. Reprinted from (Han et al., 2019) with permission.

with that of the three-dimensional mean curvature κ_m. The curvatures are computed on the center plane (z = 0 mm) of the detection volume. These *PDF*s are evaluated based on 2430 individual reconstructed CH$_2$O structures, and the same data are used for all the following analysis. In general, the *PDF*s reveal similar trends for 2D and 3D curvatures with peak values near 0 mm^{-1}. However, narrower PDFs for the 2D curvatures indicate that the lack of out-of-plane information leads to significant underestimation of the full 3D curvature. The 3D curvatures provide a more complete measure of the flame topology and are used in the following analysis.

As discussed in section 3, the inner CH$_2$O surface is curved differently than the outer surface. In Fig. 9, the *PDF*s of 3D mean curvature κ_m for the inner and outer surfaces are compared. The highest probabilites occur at small curvatures around

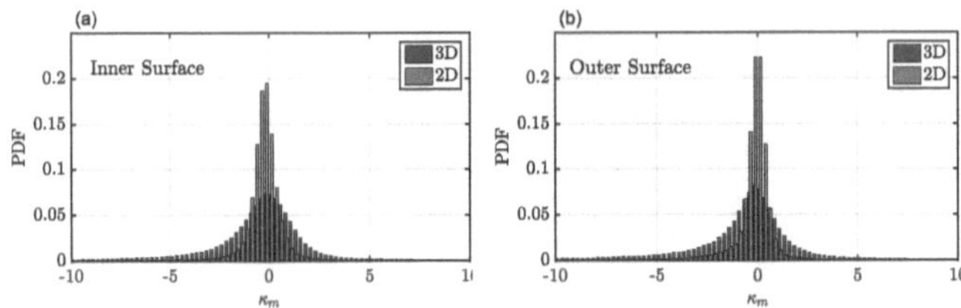

Fig. 8: The *PDF* distributions of mean curvature κ_m (in mm^{-1}) calculated from two-dimensional and three-dimensional data for (a) inner and (b) outer CH$_2$O surfaces. *PDF* bins = 0.25 mm^{-1}.

0 mm^{-1} and are slightly shifted to negative values of κ_m. The PDFs decrease monotonically towards larger magnitudes of κ_m. Figure 9 shows that the outer layer has a narrower distribution and higher probabilities at smaller mean curvatures. The spatial resolution of 3D reconstructions is determined by the z-discretization of laser sheets, which is the main factor in determining the uncertainty of the curvature calculations. Based on the average z-discretization of 250 μm, the maximum fully-resolved curvature is approximately 4 mm^{-1}. Most of the curvatures in the κ_m-PDFs for the inner and outer surfaces are within this resolution limit and are thus considered to be reliable. Additionally, the curvature of the largest vortical structures κ_{vor} (gray dashed lines) was calculated (gray dashed lines) based on the estimation of the integral length scale of $\sim$1.75 mm, resulting in $\kappa_{vor} \approx 1/(0.5 \times 1.75) \approx$1.14 mm^{-1}. Both larger and smaller surface curvatures are observed with respect to this length scale. The reduction in curvature from the inner to outer surfaces is attributed to the heat release in the CH$_2$O region where a diffusion flame resides. Turbulent eddies that traverse through the CH$_2$O region are weakened by dilatation and increasing viscositiy, resulting in a less wrinkled CH$_2$O outer surface and a narrower curvature *PDF* distribution. Moreover, the shear layer between the jet and coflow is probably located closer to the inner surface, which would lead to stronger flame stretch effects.

4.2 CH$_2$O surface shape analysis

The structure of turbulent flames is of interest for understanding turbulence-flame interactions. For example, the local flame curvature can impact the effects of transport on turbulent non-premixed flames (Han et al., 2019). With the 3D reconstructed CH2O surface, the surface curvature can be analyzed using the $\kappa_m - \kappa_g$ diagram to determine the distribution of different flame topologies. Figure 10(a) and (b) show the *JPDF* of $\kappa_m - \kappa_g$ for the inner and outer surfaces, respectively. The *JPDF*s for inner and outer surfaces have similar distributions, and the convex ($\kappa_m > 0$) structures have a slightly lower probability compared to the concave ($\kappa_m < 0$) structures. The distributions are also slightly skewed towards the saddle concave zone in the lower

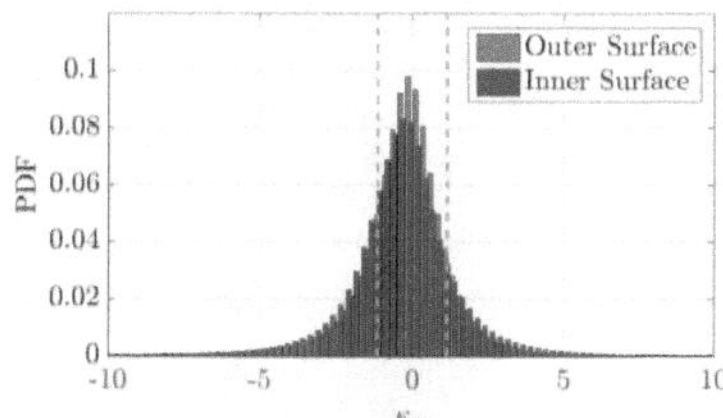

Fig. 9: The *PDF* distributions of 3D mean curvature κ_m (in mm^{-1})for inner and outer CH$_2$O surfaces. The gray dashed lines indicate the estimated curvature of largest vortical structures in the flow field. *PDF* bins $= 0.25$ mm^{-1}.

left quadrant, and which is more pronounced for outer surfaces. The saddle structures ($\kappa_g < 0$) have a higher probability than elliptic structures ($\kappa_g > 0$). For both surfaces, the highest probabilities were observed in the region of $\kappa_g \approx 0$, which indicates that the sheet-like ($\kappa_g \approx 0$ and $\kappa_m \approx 0$) and cylinder-like ($\kappa_g \not\approx 0$ and $\kappa_m \approx 0$) structures are preferential in the reaction zone of CH$_2$O.

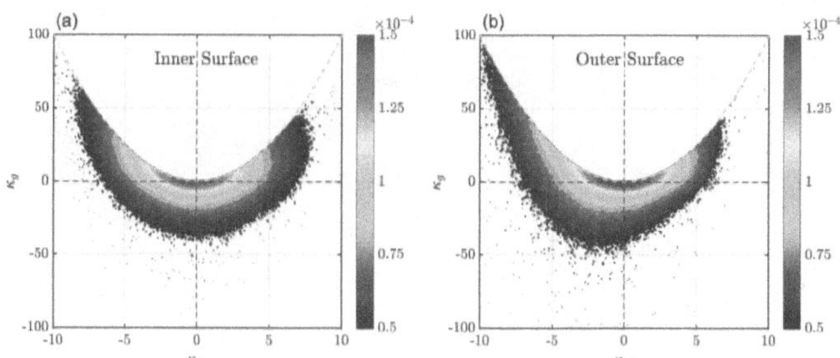

Fig. 10: The *JPDF* distribution of κ_m (in mm^{-1}) and κ_g (in mm^{-2})for (a) inner and (b) outer CH$_2$O surfaces.

In addition to the $\kappa_m - \kappa_g$ method, the shape can be investigated using principal curvatures. The two principal curvatures, κ_1 and κ_2, were computed according to Equation 5, and their *JPDF*s are plotted in Fig. 11. Because $\kappa_1 > \kappa_2$, all of the sample points are below the diagonal $\kappa_1 = \kappa_2$. In the $\kappa_1 - \kappa_2$ *JPDF*, similar to the $\kappa_m - \kappa_g$ space, the following characteristic zones are identified by the circled numbers in Fig. 11: (1) elliptic convex , (2) saddle convex, (3) saddle concave and (4) elliptic concave. The cylindrical structures can be identified along the axis of $\kappa_1 = 0$ and the axis of $\kappa_2 = 0$. The $\kappa_1 - \kappa_2$ diagram reveals similar trends to the $\kappa_m - \kappa_g$ regime. The saddle and cylindrical structures are dominant. Furthermore, the points on the outer surface are more confined within the range of small curvatures compared to those on the inner surface. The decrease of surface wrinkling is probably attributable to the turbulence degradation through the CH_2O region due to the temperature increase.

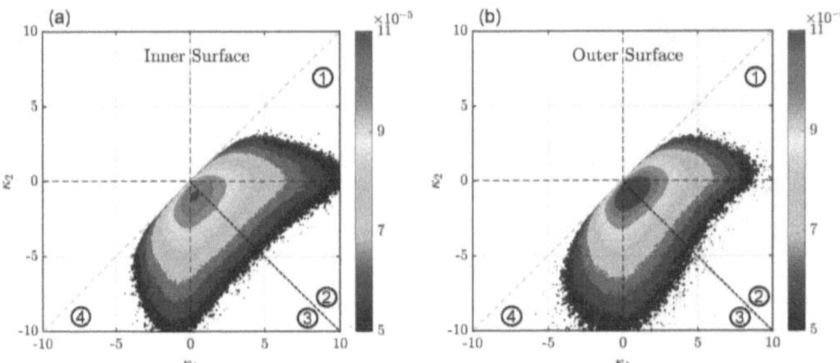

Fig. 11: The *JPDF* distribution of principal curvatures κ_1 (in mm^{-1}) and κ_2 (in mm^{-1}) for (a) inner and (b) outer CH_2O surfaces.

The characteristic shapes of three-dimensional CH_2O surfaces were further determined by the dimensionless shape factor sf, which is defined as the ratio between the principal curvatures κ_1 and κ_2:

$$sf = \frac{\kappa_1}{\kappa_2}, with \mid \kappa_1 \mid < \mid \kappa_2 \mid \tag{8}$$

and

$$sf = \frac{\kappa_2}{\kappa_1}, with \mid \kappa_1 \mid > \mid \kappa_2 \mid \tag{9}$$

Here, shape factor values sf are restricted to the range of [-1, 1]. The positive (or negative) sign of the shape factor indicates an elliptic (or saddle) surface, and $sf = 0$ signifies a cylindrical surface. In particular, $sf = 1$ indicates a spherical surface shape. As shown in Fig. 12, the shape factor *PDFs* peak at $sf \approx 0$, and no obvious preferential value other than $sf = 0$ is observed. The distributions from the inner and outer surfaces show essentially the same trends in which the probabilities of the saddle and cylindrical shapes on the surface are markedly higher than that of elliptic surfaces. This conclusion is consistent with the observation in the $\kappa_m - \kappa_g$ distributions. The sf PDFs for the inner and outer surfaces nearly overlap, indicating that the shape factor remains self-similar in these regions, despite the differences in distributions of the principal curvatures shown in Fig. 11.

Comparing the results of the $\kappa_m - \kappa_g$ and sf PDFs, similar conclusions can be made on the CH_2O surface shape characteristics. While the sf PDFs reveal the probabilities of saddle, cylindrical and elliptic shapes, $\kappa_m - \kappa_g$ indicates additionally the curvature direction with respect to the direction of scalar gradients. This makes the $\kappa_m - \kappa_g$ analysis preferred for future studies.

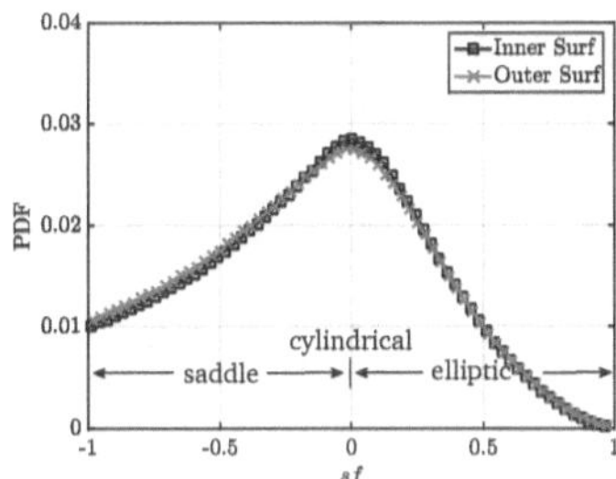

Fig. 12: The *PDF* of shape factors sf evaluated for inner and outer CH$_2$O surfaces.

4.3 Time-resolved volumetric flame topology

The 3D reconstruction of CH$_2$O-LIF images acquired from a reconstructed 5 ms burst of laser pulses provided approximately 50 consecutive instances of 3D CH$_2$O structures. Figure 13 shows a sequence of nine consecutive CH$_2$O-LIF measurements acquired at 10 kHz, illustrating the temporal and spatial evolution of the turbulent lifted flame near its base. The occurrence of an isolated CH$_2$O pocket is recorded starting from $t_0 + 200\,\mu$s. This pocket grows and propagates until it merges with the main branch at $t_0 + 600\,\mu$s. This observation could suggest the occurrence of a local auto-ignition event or simply the interaction with other regions of CH$_2$O due to out-of-plane motion. Auto-ignition in the current flame could be expected if the turbulent perturbation of the temperature field and transport of radicals to upstream locations favorably facilitates the low-temperature chemistry (Minamoto and Chen, 2016) of the DME mixture. An auto-ignition event was observed in similar turbulent lifted DME flames in a hot co-flow (Macfarlane et al., 2018). Generally, high-speed 3D imaging has the advantage of resolving ambiguities of out-of-plane motion that arise in high-speed 2D imaging. However, the isolated CH$_2$O pocket observed at $t_0 + 200\,\mu$s is on the edge of the probe volume, and therefore the effect of out-of-plane movement cannot be excluded due to the finite depth of the current 3D CH$_2$O measurements.

5 Conclusions

We have demonstrated a new capability for high-speed 3D CH$_2$O LIF measurements in a partially premixed lifted turbulent DME/air jet flame using an AOD scanning

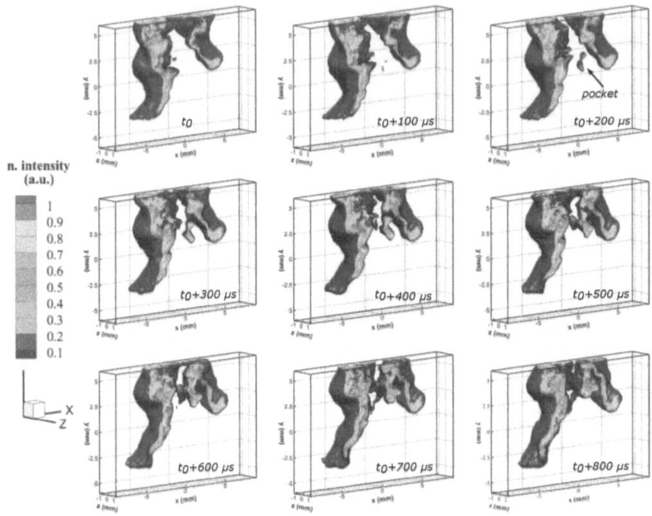

Fig. 13: Time sequence of 3D CH_2O-LIF signal reconstructions. The observation of an isolated pocket originated from $t_0+200\,\mu s$ is pointed out.

system combined with a 100 kHz pulse-burst laser. The stable and precise laser deflection in combination with a relatively simple optical setup enables reliable 3D reconstruction from the parallel laser-sheet illumination with high accuracy and precision. Consequently, a 10 kHz volumetric imaging system of CH_2O-LIF was successfully demonstrated with a detection volume of $17.3 \times 11.9 \times 2.3\,mm^3$ and a signal-to-noise ratio of up to 15. The average in-plane and out-of-plane spatial resolution was 175 µm and 250 µm, respectively, which exceeded the spatial resolution for state-of-the-art volumetric illumination-based 3D imaging techniques.

The 3D measurements were used to investigate the structural topology of CH_2O iso-surfaces. For this purpose, reliable 3D flame surface detection was achieved using adaptive intensity thresholding based on a gradient percentile method. Furthermore, systematic analysis of the Gaussian curvature κ_g and mean curvature κ_m was performed for both the inner and outer CH_2O surfaces. The statistical analysis revealed that the saddle and cylindrical structures are dominant, and curvatures on the outer surfaces have a narrower distribution than those on the inner surface, which is attributed to weakening turbulence eddies as a result of the heat release. The analysis of the principal curvatures $\kappa_1 - \kappa_2$ *JPDF* and shape factor further confirmed that the surface morphology has a greater probability of having a saddle shape than an elliptic shape. The topology statistics on the inner and outer flame surfaces showed self-similarity despite differences in the widths of the curvature distributions. To demonstrate the capability of high-speed 3D CH$_2$O-LIF measurements, the large-scale movement and deformation of the flame structures was tracked in space and time. In summary, the present results demonstrated a technique for reliable 4D scalar visualization in a turbulent reacting flow. The coupling of this technique with recent advances in 4D velocimetry techniques will provide unique possibilities for a deepened understanding of complex turbulence-chemistry interactions in turbulent reacting flows.

Acknowledgements The authors thank the Deutsche Forschungsgemeinschaft (DFG, German Research Founda-tion) – Projektnummer 215035359 – TRR 129 for its support through CRC/Transregio 129 "Oxy-flame: development of methods and models to describe solid fuel reactions within an oxy-fuel atmosphere." A. Dreizler is grateful for support by the Gottfried Wilhelm Leibniz Program of the Deutsche Forschungs-gemeinschaft. The support of the U.S. Department of Energy, Office of Basic Energy Sciences, Division of Chemical Sciences, Geosciences, and Biosciences is gratefully acknowledged. Sandia National Laborato-ries is a multimission laboratory managed and operated by National Technology and Engineering Solutions of Sandia, LLC., a wholly owned subsidiary of Honeywell International, Inc., for the U.S. Department of Energy's National Nuclear Security Administration under contract DE-NA-0003525.The views expressed in this article do not necessarily represent the views of the U.S. Department of Energy or the United States Government.

References

Baum E, Peterson B, Surmann C, Michaelis D, Böhm B, Dreizler A (2013) Investigation of the 3D flow field in an IC engine using tomographic PIV. Proceedings of the Combustion Institute 34(2):2903–2910, DOI 10.1016/j.proci.2012.06.123

Bode J, Schorr J, Krüger C, Dreizler A, Böhm B (2017) Influence of three-dimensional in-cylinder flows on cycle-to-cycle variations in a fired stratified DISI engine measured by time-resolved dual-plane PIV. Proceedings of the Combustion Institute 36(3):3477–3485, DOI 10.1016/j.proci.2016.07.106

Bode J, Schorr J, Krüger C, Dreizler A, Böhm B (2019) Influence of the in-cylinder flow on cycle-to-cycle variations in lean combustion DISI engines measured by high-speed scanning-PIV. Proceedings of the Combustion Institute 37(4):4929–4936, DOI 10.1016/j.proci.2018.07.021

Boxx I, Heeger C, Gordon R, Böhm B, Aigner M, Dreizler A, Meier W (2009) Simultaneous three-component PIV/OH-PLIF measurements of a turbulent lifted, C3H8-Argon jet diffusion flame at 1.5 kHz repetition rate. Proceedings of the Combustion Institute 32(1):905–912, DOI 10.1016/j.proci.2008.06.023

Buckmaster J (2002) Edge-flames. Progress in Energy and Combustion Science 28(5):435–475, DOI 10.1016/S0360-1285(02)00008-4

Canny J (1986) A Computational Approach to Edge Detection. IEEE Transactions on Pattern Analysis and Machine Intelligence PAMI-8(6):679–698, DOI 10.1109/TPAMI.1986.4767851

Coriton B, Steinberg AM, Frank JH (2014) High-speed tomographic PIV and OH PLIF measurements in turbulent reactive flows. Experiments in Fluids 55(6):261, DOI 10.1007/s00348-014-1743-3

Frank JH, Lyons KM, Long MB (1991) Technique for three-dimensional measurements of the time development of turbulent flames. Optics letters 16(12):958–960, DOI 10.1364/OL.16.000958

Gabet KN, Shen H, Patton RA, Fuest F, Sutton JA (2013) A comparison of turbulent dimethyl ether and methane non-premixed flame structure. Proceedings of the Combustion Institute 34(1):1447–1454, DOI 10.1016/j.proci.2012.06.183

Gautam T (1984) Lift-off Heights and Visible Lengths of Vertical Turbulent Jet Diffusion Flames in Still Air. Combustion Science and Technology 41(1-2):17–29, DOI 10.1080/00102208408923819

Gordon RL, Boxx I, Carter C, Dreizler A, Meier W (2012) Lifted Diffusion Flame Stabilisation: Conditional Analysis of Multi-Parameter High-Repetition Rate Diagnostics at the Flame Base. Flow, Turbulence and Combustion 88(4):503–527, DOI 10.1007/s10494-011-9365-9

Halls BR, Thul DJ, Michaelis D, Roy S, Meyer TR, Gord JR (2016) Single-shot, volumetrically illuminated, three-dimensional, tomographic laser-induced-fluorescence imaging in a gaseous free jet. Optics express 24(9):10040–10049, DOI 10.1364/OE.24.010040

Halls BR, Gord JR, Meyer TR, Thul DJ, Slipchenko M, Roy S (2017a) 20-kHz-rate three-dimensional tomographic imaging of the concentration field in a turbulent jet. Proceedings of the Combustion Institute 36(3):4611–4618, DOI 10.1016/j.proci.2016.07.007

Halls BR, Hsu PS, Jiang N, Legge ES, Felver JJ, Slipchenko MN, Roy S, Meyer TR, Gord JR (2017b) kHz-rate four-dimensional fluorescence tomography using an ultraviolet-tunable narrowband burst-mode optical parametric oscillator. Optica 4(8):897, DOI 10.1364/OPTICA.4.000897

Han W, Scholtissek A, Dietzsch F, Jahanbakhshi R, Hasse C (2019) Influence of flow topology and scalar structure on flame-tangential diffusion in turbulent non-premixed combustion. Combustion and Flame 206:21–36, DOI 10.1016/j.combustflame.2019.04.038

Karami S, Hawkes ER, Talei M, Chen JH (2016) Edge flame structure in a turbulent lifted flame: A direct numerical simulation study. Combustion and Flame 169:110–128, DOI 10.1016/j.combustflame.2016.03.006

Kioni PN, Rogg B, Bray K, Liñán A (1993) Flame spread in laminar mixing layers: The triple flame. Combustion and Flame 95(3):276–290, DOI 10.1016/0010-2180(93)90132-M

Kychakoff G, Paul PH, van Cruyningen I, Hanson RK (1987) Movies and 3-D images of flowfields using planar laser-induced fluorescence. Applied optics 26(13):2498–2500, DOI 10.1364/AO.26.002498

Lawn CJ (2009) Lifted flames on fuel jets in co-flowing air. Progress in Energy and Combustion Science 35(1):1–30, DOI 10.1016/j.pecs.2008.06.003

Li T, Pareja J, Becker L, Heddrich W, Dreizler A, Böhm B (2017) Quasi-4D laser diagnostics using an acousto-optic deflector scanning system. Applied Physics B 123(3):1243, DOI 10.1007/s00340-017-6663-5

Li T, Pareja J, Fuest F, Schütte M, Zhou Y, Dreizler A, Böhm B (2018) Tomographic imaging of OH laser-induced fluorescence in laminar and turbulent jet flames. Measurement Science and Technology 29(1):15206, DOI 10.1088/1361-

6501/aa938a

Lyons KM (2007) Toward an understanding of the stabilization mechanisms of lifted turbulent jet flames: Experiments. Progress in Energy and Combustion Science 33(2):211–231, DOI 10.1016/j.pecs.2006.11.001

Ma L, Lei Q, Capil T, Hammack SD, Carter CD (2017a) Direct comparison of two-dimensional and three-dimensional laser-induced fluorescence measurements on highly turbulent flames. Optics letters 42(2):267–270, DOI 10.1364/OL.42.000267

Ma L, Lei Q, Ikeda J, Xu W, Wu Y, Carter CD (2017b) Single-shot 3D flame diagnostic based on volumetric laser induced fluorescence (VLIF). Proceedings of the Combustion Institute 36(3):4575–4583, DOI 10.1016/j.proci.2016.07.050

Macfarlane AR, Dunn M, Juddoo M, Masri A (2018) The evolution of autoignition kernels in turbulent flames of dimethyl ether. Combustion and Flame 197:182–196, DOI 10.1016/j.combustflame.2018.07.022

Miller VA, Troutman VA, Hanson RK (2014) Near-kHz 3D tracer-based LIF imaging of a co-flow jet using toluene. Measurement Science and Technology 25(7):75403, DOI 10.1088/0957-0233/25/7/075403

Minamoto Y, Chen JH (2016) DNS of a turbulent lifted DME jet flame. Combustion and Flame 169:38–50, DOI 10.1016/j.combustflame.2016.04.007

Olofsson J, Richter M, Aldén M, Augé M (2006) Development of high temporally and spatially (three-dimensional) resolved formaldehyde measurements in combustion environments. Review of Scientific Instruments 77(1):13104, DOI 10.1063/1.2165569

Pareja J, Johchi A, Li T, Dreizler A, Böhm B (2019) A study of the spatial and temporal evolution of auto-ignition kernels using time-resolved tomographic OH-LIF. Proceedings of the Combustion Institute 37(2):1321–1328, DOI 10.1016/j.proci.2018.06.028

Patrie BJ (1994) Instantaneous three-dimensional flow visualization by rapid acquisition of multiple planar flow images. Optical Engineering 33(3):975, DOI 10.1117/12.160888

Peters N (2000) Turbulent combustion. Cambridge monographs on mechanics, Cambridge University Press, Cambridge, DOI 10.1017/CBO9780511612701, URL https://doi.org/10.1017/CBO9780511612701

Peterson B, Baum E, Böhm B, Dreizler A (2015) Early flame propagation in a spark-ignition engine measured with quasi 4D-diagnostics. Proceedings of the Combustion Institute 35(3):3829–3837, DOI 10.1016/j.proci.2014.05.131

Peterson B, Baum E, Ding CP, Michaelis D, Dreizler A, Böhm B (2017) Assessment and application of tomographic PIV for the spray-induced flow in an IC engine. Proceedings of the Combustion Institute 36(3):3467–3475, DOI 10.1016/j.proci.2016.06.114

Römer G, Bechtold P (2014) Electro-optic and Acousto-optic Laser Beam Scanners. Physics Procedia 56:29–39, DOI 10.1016/j.phpro.2014.08.092

Shimura M, Ueda T, Choi GM, Tanahashi M, Miyauchi T (2011) Simultaneous dual-plane CH PLIF, single-plane OH PLIF and dual-plane stereoscopic PIV measurements in methane-air turbulent premixed flames. Proceedings of the Combustion Institute 33(1):775–782, DOI 10.1016/j.proci.2010.05.026

Trunk PJ, Boxx I, Heeger C, Meier W, Böhm B, Dreizler A (2013) Premixed flame propagation in turbulent flow by means of stereoscopic PIV and dual-plane OH-PLIF at sustained kHz repetition rates. Proceedings of the Combustion Institute 34(2):3565–3572, DOI 10.1016/j.proci.2012.06.025

Weinkauff J, Greifenstein M, Dreizler A, Böhm B (2015) Time resolved three-dimensional flamebase imaging of a lifted jet flame by laser scanning. Measurement Science and Technology 26(10):105201, DOI 10.1088/0957-0233/26/10/105201

Wellander R, Richter M, Aldén M (2011) Time resolved, 3D imaging (4D) of two phase flow at a repetition rate of 1 kHz. Optics express 19(22):21508–21514, DOI 10.1364/OE.19.021508

Wellander R, Richter M, Aldén M (2014) Time-resolved (kHz) 3D imaging of OH PLIF in a flame. Experiments in Fluids 55(6):579, DOI 10.1007/s00348-014-1764-y

Wu Y, Xu W, Lei Q, Ma L (2015) Single-shot volumetric laser induced fluorescence (VLIF) measurements in turbulent flows seeded with iodine. Optics express 23(26):33408–33418, DOI 10.1364/OE.23.033408

Yip B, Schmitt RL, Long MB (1988) Instantaneous three-dimensional concentration measurements in turbulent jets and flames. Optics letters 13(2):96, DOI 10.1364/OL.13.000096

Zhou B, Frank JH (2019) Effects of heat release and imposed bulk strain on alignment of strain rate eigenvectors in turbulent premixed flames. Combustion and Flame 201:290–300, DOI 10.1016/j.combustflame.2018.12.016

Appendix C (Paper III):
Combined scanning LIF and tomographic PIV measurements

Simultaneous 10 kHz three-dimensional CH_2O and tomographic PIV measurements in a lifted partially-premixed jet flame

Bo Zhou[1,2], Tao Li[3], Jonathan H. Frank[1], Andreas Dreizler[3], Benjamin Böhm[3]

[1] *Combustion Research Facility, Sandia National Laboratories, Livermore, CA 94551, USA*
[2] *Department of Mechanics and Aerospace Engineering, Southern University of Science and Technology, Shenzhen 518055, P.R. China*
[3] *Reaktive Strömung und Messtechnik, Technische Universität Darmstadt, Otto-Berndt-Str. 3, 64287 Darmstadt, Germany*

Corresponding Author:

Tao Li

Email: tao.li@rsm.tu-darmstadt.de

Colloquium: Diagnostics

Total word count (method 1):

Main text: 526(Intro) + 956(Exp) + 1902(R&D) + 335(Con) + 117(Ac) = 3836

References: (30+2)*2.3*7.6=559

Equation.1: (1+2)*7.6=23

Figure 1: (58.5+10)*2.2*2+43=344

Figure 2: (60.1+10)*2.2*2+33=341

Figure 3: (70.6+10)*2.2*1+44=221

Figure 4: (50.8+10)*2.2*1+18=152

Figure 5: (79.3 +10)*2.2*1+32=228

Figure 6: (77.8 +10)*2.2*1+39=232

Total word count: 5936

Color Reproduction: Color figures in electronic version of manuscript.

Simultaneous 10 kHz three-dimensional CH₂O and tomographic PIV measurements in a lifted partially-premixed jet flame

Bo Zhou[1,2], Tao Li[3], Jonathan H. Frank[1], Andreas Dreizler[3], Benjamin Böhm[3]

[1] *Combustion Research Facility, Sandia National Laboratories, Livermore, CA 94551, USA*

[2] *Department of Mechanics and Aerospace Engineering, Southern University of Science and Technology, Shenzhen 518055, P.R. China*

[3] *Reaktive Strömung und Messtechnik, Technische Universität Darmstadt, Otto-Berndt-Str. 3, 64287 Darmstadt, Germany*

Keywords:

Turbulent lifted jet flame, High-speed 3D CH₂O-LIF, High-speed TPIV, 3D displacement speed, Lagrange particle trajectory

1. Introduction

Turbulent lifted flames stabilize downstream of the fuel nozzle within a three-dimensional (3D) turbulent flow without a physical flame stabilizer. Such flame structures are of relevance for industrial burners and commercial boilers but also exist in the near field of fuel jets in diesel engines and gas turbines. The flame stabilization mechanism of turbulent lifted jet flames has been the focus of experimental and numerical investigations [1-4]. However, the interpretation of such transient three-dimensional events from planar and temporally uncorrelated experimental data is difficult. The application of high-speed (HS) planar multi-parameter diagnostics has supported the interpretation of the underlying processes. Upatnieks et al. [5] applied high-speed particle image velocimetry (HS-PIV) and flame imaging from evaporating droplets to lifted flames. Little correlation was found between the flow field and flame propagation. In this context, the influence of out-of-plane flow velocities has been discussed [6]. In a subsequent study, all three velocity components were measured using a combination of HS-stereo-PIV, which was combined with HS-planar laser induced fluorescence (PLIF) of OH [7]. Flame motion statistics were evaluated based solely on data with negligible out-of-plane fluid motion. The flame displacement speed (s_d) relative to the flow was a few times the laminar flame speed. This is in agreement with other studies, such as those based on PIV and dual-shot CH-PLIF [8] and combined stereo-PIV and dual-plane OH-PLIF [9]. Ultimately, determination of s_d requires simultaneous high-speed 3D scalar/flow field measurements that provide complete information on flame location and orientation together with local velocity fields. Additionally, such measurements would provide a unique experimental assessment of Lagrangian analysis in flames, which has been shown to be valuable from numerical studies and provides a new perspective in studying turbulence-flame interactions [10].

For high-speed volumetric flow measurements, tomographic PIV (TPIV) in turbulent reacting flow has been available through recent developments [11]. HS-volumetric scalar measurements are more challenging. HS-volumetric measurements for OH [12], fuel-tracer [13] and soot [14] have been shown to be feasible using volumetric laser illumination but demand high laser flux and suffer from low spatial resolution. Recently, multiple laser sheet approaches using an acousto-optic deflector (AOD) have been proposed, where a single laser sheet is scanned rapidly through a measurement volume. In comparison to previous scanning concepts using a rotating mirror [15, 16], an AOD does not contain any moving parts and thus delivers advantages over mechanical scanners in terms of scan frequency, accuracy, precision and spatial resolution [17]. The capabilities of AODs for laser combustion diagnostics have been demonstrated for 3D-flame visualizations of a turbulent lifted jet flame by means of Mie-scattering [18] and LIF-imaging [19].

This paper demonstrates simultaneous time-resolved volumetric measurements of the flow and the flame at the base of a turbulent lifted partially premixed DME/air flame. Combined high-speed TPIV and 3D formaldehyde (CH_2O) PLIF were realized using the AOD concept. In the following, the experimental setup, diagnostic details and data processing are provided first. Taking advantage of this cinematographic volumetric flow-scalar measurement, we demonstrate its capability for Lagrangian analysis and s_d quantification on the CH_2O surface of the flame base. Results provide insights into the stabilization mechanism of the lifted turbulent jet flame of the present study.

2. Experiment

2.1 Partially premixed DME/air jet flame

Figure 1a shows the experimental setup used to study a partially premixed lifted turbulent DME/air jet flame. The burner consisted of an air co-flow of 76 mm inner diameter (ID) and a central nozzle of 2.5 mm ID supplied with the DME/air mixture having an equivalence ratio of 9.5. The co-flow bulk velocity was 0.2 m/s, and the jet bulk velocity at the nozzle exit was 17 m/s, resulting in a Reynolds number of approximately 4500. The lifted DME/air jet flame was stabilized downstream of the jet exit with a mean lift-off height of about 25 mm.

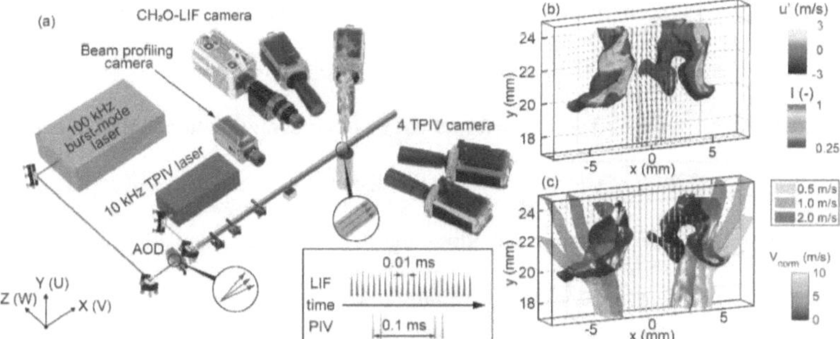

Figure 1. (a) Experimental setup. (b) Instantaneous CH_2O-LIF signal with the fluctuating velocity field colored by the magnitudes of axial velocity fluctuation v. (c) The extracted CH_2O inner (blue) and outer (red) surfaces in the mean velocity field with the mean velocity iso-surfaces V_{norm} (gray).

2.2. CH₂O-LIF Scanning System

The CH_2O-LIF was excited by the third harmonic ($\lambda = 355$ nm) of the pulse-burst laser, which provided a pulse train at 100 kHz with 5 ms duration. A water-cooled quartz crystal (D1340, ISOMET) combined with a tunable RF-driver (RFA333, ISOMET) were applied to sweep the laser beam. Details of laser scanning using AOD are described in previous work [18, 19]. The laser beam had a nearly constant pulse energy of 2 mJ and was formed into a 25-mm high laser sheet with a thickness of 100 μm at the focus near the jet centerline. The laser pulse was synchronized with the AOD such that every scan cycle included 10 laser pulses with a cycle frequency of 10 kHz. The parallelization of multiple laser sheets was monitored on a beam-profiling camera. The scan depth was 2.3 mm with a maximum deviation of <100 μm within ±15 mm along the x-axis.

The CH_2O-LIF signal was collected by a CMOS camera (Fastcam SA-X2) combined with a lens-coupled intensifier (HICATT). The camera was equipped with a 58 mm lens (Nikkor, f/2.8) and had a depth of field (DOF) of >3 mm. A 500 nm short-pass filter and a 355 nm laser-line notch filter suppressed flame luminosity and blocked scattered laser light. The field of view (FOV) was 17.3 (width) ×11.9 (height) mm^2 with a projected pixel size of 45 μm. A signal-to-noise ratio (SNR) of 15 was achieved for CH_2O-LIF measurements. The in-plane resolution was approximately 175 μm as evaluated using a USAF resolution target with a relative modulation depth of 0.33. The out-of-plane resolution of 250 μm was restricted by the distance between laser-sheet planes.

The pulse-to-pulse energy fluctuation and the in-plane intensity inhomogeneity for individual laser sheets were accounted for by simultaneously imaging biacetyl LIF signal from a laminar jet with an air/biacetyl ($C_4H_6O_2$) vapor mixture. The biacetyl LIF signal (also excited at λ=355 nm) of each laser pulse was imaged onto a second intensified CMOS camera at 100 kHz with a FOV of 3.6 (width)×14.3 (height) mm^2. The camera was equipped with a 58 mm lens (Nikkor, f/2.8) and a 355 nm laser-line notch filter. The laser-sheet intensity distribution was evaluated by averaging the biacetyl-LIF image along the beam-propagation direction.

2.3. Tomographic PIV

Simultaneous 10 kHz TPIV measurements were performed using a diode-pumped dual-head Nd:YAG laser with a pulse energy of 5 mJ. A probe volume of 24.7×14.1×2.3 mm^3

overlapped with the scanning CH_2O-LIF measurement at the burner centerline. As shown in the inset of Fig. 1a, each TPIV pulse pair (Δt=15 μs) was centered in a 100-μs LIF scan cycle consisting of 10 laser pulses. Both the air coflow and DME/air jet were seeded with 0.3-μm aluminum-oxide particles for flow tracking. A pair of CMOS cameras, equipped with macro lenses (f=180 mm, DOV>3mm) on a Scheimpflug mount, were positioned on each side of the laser path in the forward scattering direction at angles of 20° and 45° with respect to the z-axis. The TPIV cameras were operated at 20 kHz in a frame straddling mode with a detection area of 1280×800 pixel^2 and a projected pixel height of 19 μm. The TPIV data was evaluated using a multiplicative algebraic reconstruction tomography (MART) algorithm with detailed pre- and post-processing described in reference [20]. The dynamic range of the high-speed TPIV measurement was increased by repetitively evaluating the velocity field using particle images from pulse pairs with different time intervals. Various pulse pair combinations provided a range of time intervals from 15 μs to 315 μs, which accommodated the large range of velocities in the present flow and sufficiently resolved fluid motion near the flame base. The ultimate flow field was constructed from a combination of different time intervals using a velocity-weighting function which gave more weights to the flow field result with an optimal particle displacement. The 3D flow field was further refined using a 4th order Runge-Kutta algorithm [21], resulting in 240×130×23 vectors with 100 μm separation.

2.4. Volumetric reconstruction of CH_2O-LIF signal

The 3D CH_2O-LIF signal matrix consisting of 10 2D images was linearly interpolated along the z-axis such that the plane spacing along the z-axis equaled the in-plane projected pixel size of 45 μm. Figure 1b shows an instantaneous CH_2O structure represented by iso-surfaces of

CH$_2$O-LIF intensities. The coordinate system (x, y, z) is centered at the jet exit. Velocity fluctuation vectors (1 out of 4 displayed) in the plane of z = 0 are colored by magnitudes of axial velocity fluctuation u ($u = U - \overline{U}$, where $\overline{U}$ is temporally-averaged axial velocity). The CH$_2$O surface is extracted by dynamically defining an intensity threshold which corresponds to local maximum intensity gradients [22]. The CH$_2$O-LIF iso-surfaces on each side of the flame base were separated into an inner (blue) and outer surface (red) relative to the jet axis, as shown in Fig. 1c. In each z-plane, the inner and outer surfaces meet at the farthest upstream position of the flame base. The mean velocity field and the iso-contours of mean velocity magnitude, V$_{norm}$ = 0.5, 1 and 2 m/s, are shown to illustrate the relative location of the CH$_2$O structure in the velocity field.

3. Results and Discussions

Simultaneous high-speed 3D CH$_2$O-LIF and TPIV measurements enable Lagrangian analysis near the flame base (Section 3.1) and calculation of the displacement speed for the CH$_2$O surface (Section 3.2). Combining both provides insights to the stabilization mechanism of the lifted turbulent partially premixed DME/air flame. The present study demonstrates this ability using a single 5 ms imaging sequence.

3.1 Lagrangian Trajectory

In this section, we demonstrate a Lagrangian analysis with known flame locations using 4D quasi-simultaneous CH$_2$O-LIF/TPIV measurements. As shown in Fig. 2a, we start by artificially seeding massless particles (representing fluid elements) on the CH$_2$O inner surface at a time instant (t=0) in the middle of a measurement sequence, which ensures that the particle trajectories of the Lagrangian analysis intersect the CH$_2$O surface. The virtual particles are tracked forwards and backwards in time to the end and beginning of the 5 ms measurement sequence.

Based on particle locations at the beginning, the trajectories are divided into three zones delineated by mean velocity iso-contours of V_{norm}, namely $V_{norm} \leq 0.5m/s$ (zone①), $0.5m/s < V_{norm} < 2m/s$ (zone ②), and $V_{norm} > 2m/s$ (zone③). Due to the unsteady nature of turbulent flow, this zone delineation is only tentative and qualitatively discriminates different trajectory characteristics for each zone. The particle trajectories in zones ① and ② are displayed in detail in Fig.2(b-c). The mean velocities along particle trajectories in each zone are shown in Fig.3(a-c). Overall, the CH_2O structure near the flame base features a "L-shape" with the horizontal leg extending into a low-speed region residing in zones ① and ② where ambient air is abundant.

Particle trajectories in zone ① (Fig.2b) exhibit vortex-like structures slightly ahead of the CH_2O inner surface, and the trajectories are approximately parallel to the surface-normal at intersections with the inner and outer surfaces. The vortex structures observed in these Lagrangian trajectories are barely visible in the fluctuating velocity field with Eulerian coordinates shown in Fig.1b. The corresponding mean velocity profiles of the trajectories in zone ① in Fig.3a further show that the vortex-like structures correspond to a deceleration (acceleration) of the axial velocity (radial velocity) such that the axial velocity (U) became slightly negative ahead of the flame base. The mean axial velocity resumes to approximately 0.3 m/s when approaching the flame base at t=0 followed by a two-stage acceleration at t=0 and 1.2 ms, respectively. Note that the particle trajectories at t=1.2 ms fall approximately on the outer surface. Considering that zone ① corresponds to the low-speed region in close proximity to the air coflow, the observed axial flow deceleration followed by a two-stage acceleration along the particle trajectories suggests a lean premixed flame behavior. The CH$_2$O layer in this region presumably represents the broadened preheat layer of a lean premixed flame with the inner surface on the reactant side and the outer surface next to the reaction zone. This is consistent with previous studies of lifted jet flames where the high-temperature region marker, OH radicals, surrounds the outer surface of reactants [23]. The flow deceleration is likely caused by heat-release induced dilatation, forming the observed vortex structure on the particle trajectories slightly ahead of the CH$_2$O inner surface shown in Fig.2b. Once the fluid particles enter the CH$_2$O region (namely the preheat layer), particle acceleration is driven by the temperature difference across the inner and outer CH$_2$O surfaces. The trajectories across the outer surface suggest the transport of hot combustion products including OH radicals that would provide support for the flame downstream.

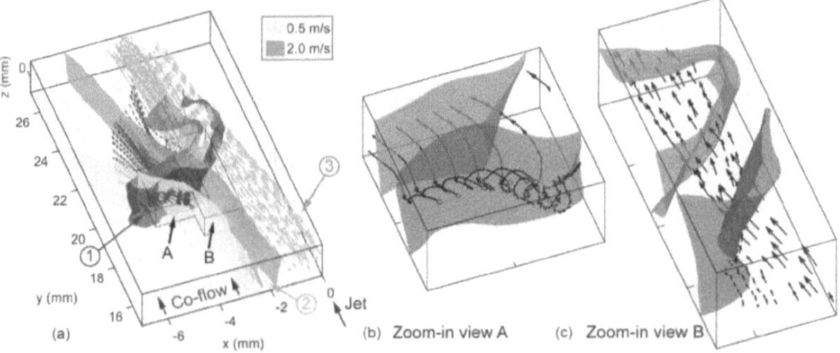

Figure 2. (a) Particle trajectories together with the instantaneous CH$_2$O iso-surface at t=0. (b-c) Zoomed-in plots of boxes A and B. Velocity iso-surfaces in gray correspond to the mean V$_{norm}$ = 0.5 and 2 m/s.

The trajectories in zone ② reside approximately in the shear layer and contain virtual particles with both low-speed and high-speed origins as shown in Fig. 2c. The mean axial velocity of the trajectories increases from approximately 1 m/s to 4 m/s upstream of the flame base (t<0) presumably due to flow mixing with the central jet. At t=0, the trajectory velocities in Fig. 3b show

no obvious evidence of flow acceleration across the surface and thereafter, suggesting absence of

strong heat release in zone ②. The trajectories in zone ③ have the highest axial velocities

originating from the central fuel jet stream, and chemical reactions in this region are significantly

suppressed due to lack of adequate oxidizer. Mean axial velocities of the trajectories experience a

decrease after crossing the CH_2O inner surface at t=0 within the hot CH_2O layer.

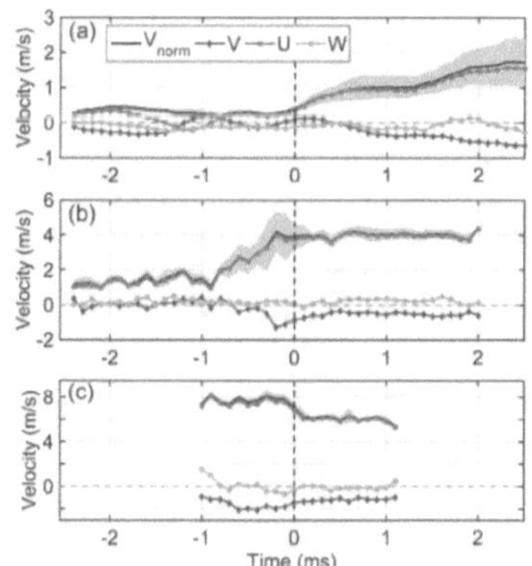

Figure 3. Mean velocity components and magnitudes along particle trajectories in (a) zone ① (b)

zone ② and (c) zone ③. All trajectories are centered at t=0, the time of intersection with the inner

CH_2O iso-surface. The gray shaded region indicates the V_{norm} range with ±1 standard deviation.

Figure 4(a–b) further quantifies residence times and spatial lengths within the CH_2O

structure for trajectories in the three zones. The residence times for trajectories in zone ① span

from 0.65 to 1.25 ms with the corresponding residence lengths ranging from 0.2 to 2.2 mm. The

residence lengths approximate the range of the broadened preheat layer thicknesses considering the overall parallel alignment between trajectories and surface normal. As shown in Fig.4(a-b), a large fraction of trajectories in zone ② and zone ③ exhibit residence times less than 0.5 ms. In comparison, a residence time of 0.4 ms is estimated for a flame speed of s_L=0.5 m/s and a reaction zone thickness of 200 μm. Too short residence times suggest incomplete reactions or the absence of reactions in zones ② and ③. Due to flow mixing in the shear layer, a small fraction of trajectories in zone ② show longer residence times up to 1.35 ms with corresponding residence lengths up to 3.7 mm.

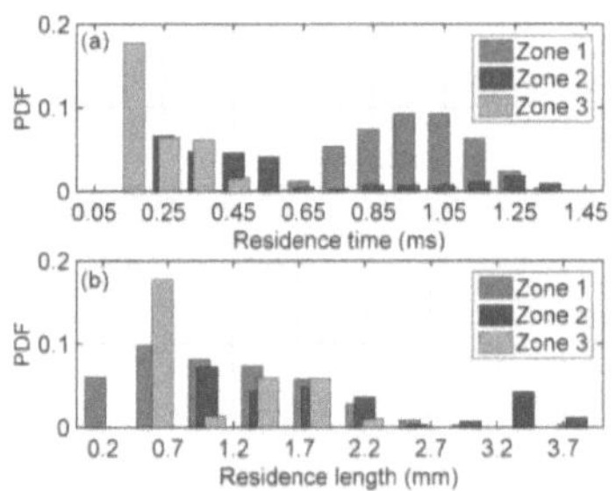

Figure 4. PDF of the (a) residence time and (b) length within CH_2O structures for trajectories in the three zones.

3.2. 3D Displacement Speed

In this section, we present quantified displacement speeds s_d of the CH_2O iso-surfaces (not to be confused with flame displacement speed of premixed flames which is often termed s_d in the literature). The absolute surface velocity $\vec{w}$ equals the sum of the local gas velocity $\vec{v_c}$ and local displacement velocity $\vec{v_d}$ expressed as $\vec{w} = \vec{v_c} + \vec{v_d} = \vec{v_c} + \vec{n} \cdot s_d$, where $\vec{n}$ denotes the CH_2O surface normal. Evaluating s_d includes two steps. As shown in Fig. 5a, the 3D CH_2O surface S_1

(red) at t_1 is first convected by the gas velocity $\vec{v_c}$ to a virtual surface S_c (blue). For this step, $\vec{v_c}$ on surface S_1 was obtained from the measured 3D velocity field. In the second step shown in Fig. 5b, s_d is determined from the displacements between surface S_c to the surface S_2 at t_1+100 μs (green) along the surface S_c normal $\vec{n}$ divided by the 100 μs time interval. The resulting s_d for the same instant of the 3D CH_2O iso-surface as in Fig. 2a is illustrated in Fig. 6a. Both positive and negative s_d are observed along the CH_2O surface with peak s_d magnitudes up to eight times the laminar burning velocity ($s_L \approx 0.45$ m/s) of a stoichiometric DME/air mixture [24]. Note that negative s_d near the points connecting the inner and outer surfaces is an artifact from the s_d calculation due to ambiguities in determining the associated points at surface S_2 to which surface S_c propagates.

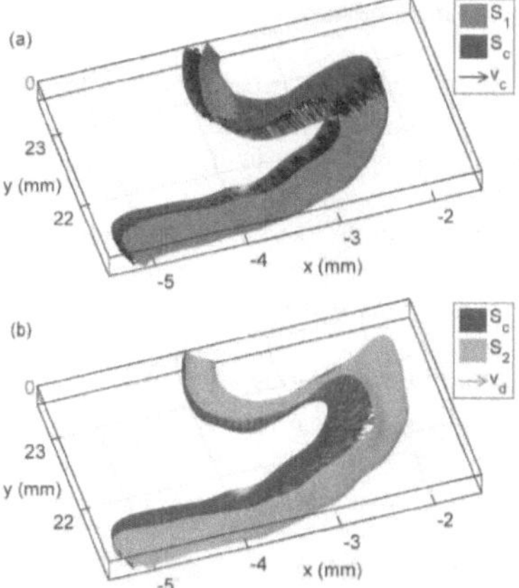

Figure 5. (a) Instantaneous CH_2O surface S_1 at t_1 (red) and convected surface S_c (blue) by local convection vector $\vec{v_c}$. (b) Instantaneous S_c and S_2 (green) at t_1+100 μs with the displacement vector $\vec{v_d}$.

To interpret the s_d distribution of the CH_2O surface, it is important to understand what the CH_2O scalar represents in the context of flame modes as local mixtures can vary dramatically in a turbulent lifted jet flame. Based on the Lagrangian analysis in the section 3.1 and a previous study of lifted jet flames [25], the lifted flame base hypothetically could be an edge flame with multiple flame modes as schematically shown in Fig. 6b for the left CH_2O branch with the delineated gray regions approximately corresponding to the three zones in the Lagrangian analysis. A stoichiometric mixture fraction contour (Z_{st}) presumably forms near the flame base within the shear layer in zone ② and mixtures at both sides of the Z_{st} are either fuel-rich (closer to jet center) or fuel-lean (further away from jet center) [26]. As discussed in section 3.1, the CH_2O layer at the flame base to the left of Z_{st} represents the broadened preheat layer of lean premixed flames, and the CH_2O layer extends further into the low-speed region of zone ① until the tip of the CH_2O layer where the mixture becomes too lean to burn. To the right of the Z_{st} line, rich-premixed or diffusion flames may develop at the outer surface as the mixture becomes richer. In reality, flame holes [27] due to local quenching may also exist, and the transition between different flame modes would not be as abrupt as shown in the schematic of Fig. 6b. Confirmation of the flame mode distribution discussed above requires simultaneous measurements of the local mixture composition and temperature, which is experimentally challenging and beyond the scope of this study.

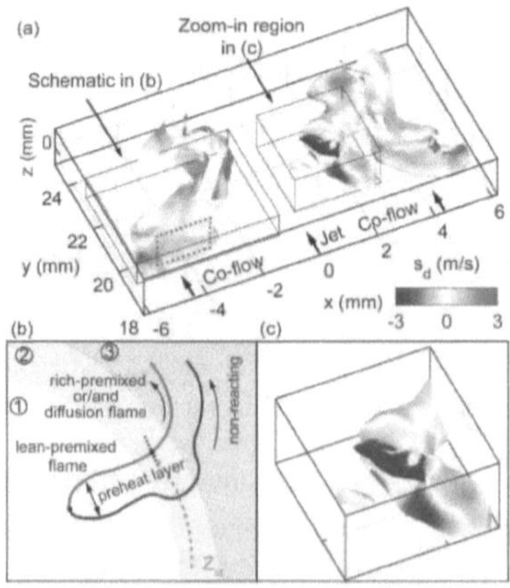

Figure 6. (a) Distribution of s_d on instantaneous 3D CH$_2$O iso-surfaces near the lifted flame base. (b) A hypothesized schematic structure corresponding to the left CH$_2$O branch near the flame base. (c) zoomed-in view of s_d distribution on local CH$_2$O structure.

It is instructive to decompose s_d into three components, namely the reaction ($s_{d,r}$), normal diffusion ($s_{d,n}$) and curvature ($s_{d,c}$) components as expressed in Eq.1 [28]:

$$s_d = \frac{\dot{\omega}}{\rho|\nabla c|} + \frac{1}{\rho|\nabla c|}\frac{\partial}{\partial n}\left(\rho D_c \frac{\partial c}{\partial n}\right) - D_c \nabla \cdot \boldsymbol{n} \qquad \text{Eq. (1)}$$

where $\dot{\omega}$ is the reaction rate, c is the scalar iso-surface with local gas density of ρ, D_c is the diffusivity and $\nabla \cdot \boldsymbol{n}$ is the curvature. According to the hypothesized flame base structure in Fig. 6b, the main chemical reaction zones are located at the outer surface either in the premixed or diffusion flame mode. The corresponding s_d is therefore a result of a reactive-diffusive balance on the surface. For a premixed flame, the resulting s_d weighted by the density variations would equal the unstrained s_L if the surface straining and curvature are negligible. When chemical reactions are at a low intensity, as for the CH$_2$O inner surface close to the fuel jet center, the corresponding s_d is governed by the $s_{d,n}$ and $s_{d,c}$ components. For the flame base in low-speed zone ① which is likely

in a premixed mode, it is of interest to observe that the overall magnitudes for s_d on the inner and outer surfaces do not exhibit a distinct difference as one otherwise would expect due to the temperature differences associated with the two preheat-layer surfaces for a laminar premixed flame. Previous studies of turbulent premixed flames have shown that the probability density functions (PDFs) of s_d for surfaces associated with different temperatures have the same most probable s_d, but the PDFs for high-temperature surfaces extend towards larger s_d values [29]. This suggests that the differences in s_d for the two preheat-layer surfaces are moderated by surface straining due to turbulence. Surface straining and thermal expansion cause changes in the scalar gradients which directly affect the $s_{d,n}$ component and the magnitudes of s_d in a highly nonlinear manner. In response to the surface straining, both positive and negative s_d have been observed with values covering multiples of the corresponding s_L [30], and negative s_d is typically correlated with high positive strains [29, 30]. In addition to surface straining, curvature also plays an important role in local s_d distribution through the $s_{d,c}$ component. Consistent with previous numerical studies [27], we observe large negative values of s_d in regions of high positive curvature (convex towards reactants) as shown in Fig. 6c. Negative values of s_d cause the cusp to recede quickly, which prevents further development of the cusp. The corresponding s_d values in zones ① and ② (highlighted in green dashed box) are mostly between 0.2 m/s and 1 m/s, which approximately matches the local flow speeds.

Many numerical studies calculate s_d at the maximum heat-release surface, but experimental determination of the heat-release surface may not always be feasible. Despite the complexity in determining s_d, its measurement in this demonstration carries a clear physical meaning that represents the relative movement of a selected scalar iso-surface. The ability to perform temporally resolved 3D measurements of s_d together with the fluid flow measurement enables future studies on the evolution of surface wrinkling as well as the effect of stretch on s_d.

4. Conclusion

This study demonstrates quasi-simultaneous high-speed 3D TPIV/CH$_2$O-LIF measurements at 10 kHz in a partially premixed lifted turbulent DME/air jet flame using an AOD scanning system combined with a 100 kHz pulse-burst laser. These high-speed volumetric diagnostics enable new experimental capabilities for Lagrangian analysis of particle trajectories with known flame locations and 3D measurements of the displacement speed (s_d).

The high-speed 3D CH$_2$O-LIF measurement reveals a "L-shape" CH$_2$O structure near the flame base with the horizontal leg of the CH$_2$O structure extending into a low-speed region abundant in air. In this region, the Lagrangian analysis of particle trajectories revealed vortex-like structures slightly ahead of the CH_2O structure, which were hardly visible in the flow field using the conventional Eulerian representation. The velocity profiles of the corresponding trajectories showed a deceleration of the axial velocity down to a slightly negative value followed by a two-stage acceleration after crossing the CH_2O structure. This observation suggests a lean-premixed flame behavior in this region. For the s_d measurements, positive and negative values of s_d were observed on the instantaneous CH_2O surface near the lifted flame base as a result of a complex interplay between reactions, surface straining, and curvature that are likely accompanied by local variations in mixture fraction and temperature. Overall, the magnitude of the s_d shows no distinct difference on the inner and outer CH_2O surfaces. Consistent with the literature, large negative values of s_d were observed in regions of high positive curvature. The maximum s_d magnitudes on the CH_2O iso-surface are about eight times the laminar burning velocity of a stoichiometric DME/air mixture, and the s_d at the flame base approximately matches the local flow speed. The observations in the present study through the Lagrangian analysis and the s_d calculation support formation of lean premixed flames near the flame base whose role in flame stabilization should be investigated in more detail. Accurate determination of different local flame modes near the flame base will require additional measurements of local mixture composition.

Acknowledgements

The support of the U.S. Department of Energy, Office of Basic Energy Sciences, Division of Chemical Sciences, Geosciences, and Biosciences is gratefully acknowledged. The authors thank the Deutsche Forschungsgemeinschaft (DFG, German Research Foundation) – Projektnummer 215035359–TRR 129 for its support through CRC/Transregio 129 "Oxy-flame:

development of methods and models to describe solid fuel reactions within an oxy-fuel atmosphere". The authors thank Prof. Xue-song Bai for insightful discussions and Mr. Erxiong Huang for technical assistance in the laboratory. Sandia National Laboratories is a multimission laboratory managed and operated by National Technology and Engineering Solutions of Sandia, LLC., a wholly owned subsidiary of Honeywell International, Inc., for the U.S. Department of Energy's National Nuclear Security Administration under contract DE-NA-0003525.

References
[1] K.M. Lyons, Toward an understanding of the stabilization mechanisms of lifted turbulent jet flames: Experiments, Prog. Energy Combust. Sci 33 (2007) 211–231.

[2] C.J. Lawn, Lifted flames on fuel jets in co-flowing air, Prog. Energy Combust. Sci 35 (2009) 1–30.

[3] S. Karami, E.R. Hawkes, M. Talei, J.H. Chen, Mechanisms of flame stabilisation at low lifted height in a turbulent lifted slot-jet flame, J. Fluid Mech. 777 (2015) 633–689.

[4] W.M. Pitts, Assessment of theories for the behavior and blowout of lifted turbulent jet diffusion flames, Prog. Energy Combust. Sci 22 (1989) 809–816.

[5] A. Upatnieks, J.F. Driscoll, C.C. Rasmussen, S.L. Ceccio, Liftoff of turbulent jet flames—assessment of edge flame and other concepts using cinema-PIV, Combust. Flame 138 (2004) 259–272.

[6] I. Boxx, C. Heeger, R. Gordon, B. Böhm, A. Dreizler, W. Meier, On the importance of temporal context in interpretation of flame discontinuities, Combust. Flame 156 (2009) 269–271.

[7] R.L. Gordon, I. Boxx, C. Carter, A. Dreizler, W. Meier, Lifted Diffusion Flame Stabilisation: Conditional Analysis of Multi-Parameter High-Repetition Rate Diagnostics at the Flame Base, Flow Turbul Combust 88 (2012) 503–527.

[8] K.A. Watson, K.M. Lyons, C.D. Carter, J.M. Donbar, Simultaneous two-shot CH planar laser-induced fluorescence and particle image velocimetry measurements in lifted CH4/air diffusion flames, Proc. Combust. Inst. 29 (2002) 1905–1912.

[9] B. Peterson, E. Baum, B. Böhm, A. Dreizler, Early flame propagation in a spark-ignition engine measured with quasi 4D-diagnostics, Proc. Combust. Inst. 35 (2015) 3829–3837.

[10] P.E. Hamlington, R. Darragh, C.A. Briner, C.A.Z. Towery, B.D. Taylor, A.Y. Poludnenko, Lagrangian analysis of high-speed turbulent premixed reacting flows: Thermochemical trajectories in hydrogen–air flames, Combust. Flame 186 (2017) 193–207.

[11] B. Coriton, A.M. Steinberg, J.H. Frank, High-speed tomographic PIV and OH PLIF measurements in turbulent reactive flows, Exp. Fluids 55 (2014) 261.

[12] T. Li, J. Pareja, F. Fuest, M. Schütte, Y. Zhou, A. Dreizler, B. Böhm, Tomographic imaging of OH laser-induced fluorescence in laminar and turbulent jet flames, Meas. Sci. Technol. 29 (2018) 15206.

[13] B.R. Halls, J.R. Gord, T.R. Meyer, D.J. Thul, M. Slipchenko, S. Roy, 20-kHz-rate three-dimensional tomographic imaging of the concentration field in a turbulent jet, P Combust Inst 36 (2017) 4611-4618.

[14] T.R. Meyer, B.R. Halls, N.B. Jiang, M.N. Slipchenko, S. Roy, J.R. Gord, High-speed, three-dimensional tomographic laser-induced incandescence imaging of soot volume fraction in turbulent flames, Opt Express 24 (2016) 29547-29555.

[15] J. Olofsson, M. Richter, M. Aldén, M. Augé, Development of high temporally and spatially (three-dimensional) resolved formaldehyde measurements in combustion environments, Rev. Sci. Instrum. 77 (2006) 13104.

[16] J. Weinkauff, M. Greifenstein, A. Dreizler, B. Böhm, Time resolved three-dimensional flamebase imaging of a lifted jet flame by laser scanning, Meas. Sci. Technol. 26 (2015) 105201.

[17] G.R.B.E. Römer, P. Bechtold, Electro-optic and Acousto-optic Laser Beam Scanners, Physics Procedia 56 (2014) 29–39.

[18] T. Li, J. Pareja, L. Becker, W. Heddrich, A. Dreizler, B. Böhm, Quasi-4D laser diagnostics using an acousto-optic deflector scanning system, Appl. Phys. B 123 (2017) 1243.

[19] T. Li, B. Zhou, J.H. Frank, A. Dreizler, B. Böhm, High-speed volumetric imaging of formaldehyde in a lifted turbulent jet flame using an acousto-optic deflector, Experiments in Fluids 61 (2020).

[20] B. Zhou, J.H. Frank, Effects of heat release and imposed bulk strain on alignment of strain rate eigenvectors in turbulent premixed flames, Combust. Flame 201 (2019) 290–300.

[21] D.L. Darmofal, R. Haimes, An Analysis of 3D Particle Path Integration Algorithms, J. Comput. Phys. 123 (1996) 182–195.

[22] J. Pareja, A. Johchi, T. Li, A. Dreizler, B. Böhm, A study of the spatial and temporal evolution of auto-ignition kernels using time-resolved tomographic OH-LIF, Proc. Combust. Inst. 37 (2019) 1321–1328.

[23] R.L. Gordon, A.R. Masri, E. Mastorakos, Heat release rate as represented by [OH] x [CH2O] and its role in autoignition, Combust Theor Model 13 (2009) 645-670.

[24] A. Mohammad, K.A. Juhany, Laminar burning velocity and flame structure of DME/methane plus air mixtures at elevated temperatures, Fuel 245 (2019) 105-114.

[25] S.H. Chung, Stabilization, propagation and instability of tribrachial triple flames, Proc. Combust. Inst. 31 (2007) 877–892.

[26] K.A. Watson, K.M. Lyons, J.M. Donbar, C.D. Carter, Simultaneous Rayleigh imaging and CH-PLIF measurements in a lifted jet diffusion flame, Combust. Flame 123 (2000) 252–265.

[27] T. Echekki, J.H. Chen, Unsteady strain rate and curvature effects in turbulent premixed methane-air flames, Combust Flame 106 (1996) 184-202.

[28] G.V. Nivarti, R.S. Cant, Stretch Rate and Displacement Speed Correlations for Increasingly Turbulent Premixed Flames, Flow Turbul Combust 102 (2019) 957-971.

[29] J.H. Chen, H.G. Im, Correlation of flame speed with stretch in turbulent premixed methane/air flames, Proc. Combust. Inst. 27 (1998) 819–826.

[30] E.R. Hawkes, J.H. Chen, Comparison of direct numerical simulation of lean premixed methane–air flames with strained laminar flame calculations, Combust. Flame 144 (2006) 112–125.

Appendix D (Paper IV):
Multi-parameter measurements of single particle combustion

were insignificant for small particles. Simulation results suggested that the local heat transfer was improved by CO_2, mainly due to the lower temperature sink close to particles and hence higher volatile release rates. As the initial ambient temperatures were similar, the introduction of CO_2 favored homogeneous ignition and slowed down the volatile consumption.

Keywords: Single particle combustion, Ignition and volatile flame, Multi-parameter laser diagnostics, Oxy-fuel combustion, Bituminous coal, Combined experimental-numerical approach

Graphical abstract

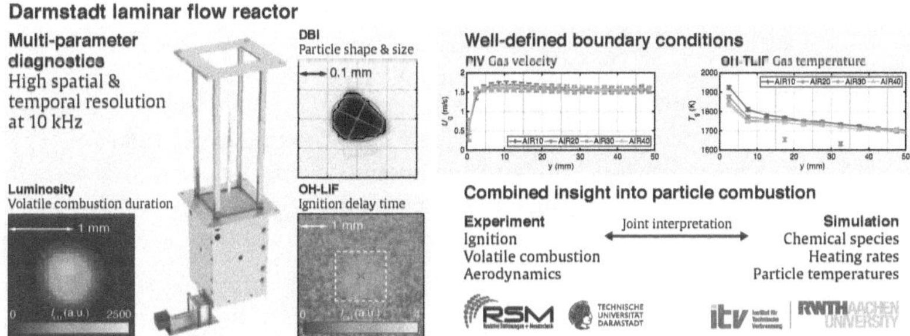

Highlights

- Single particle combustion is investigated in a laminar flow reactor with fully-measured boundary conditions using a combined experimental-numerical approach

- Multi-parameter optical diagnostics provide comprehensive data including the particle velocity, ignition delay time and volatile combustion duration

- Simulations assist to interpret the experimental observations with varying particle sizes, oxygen concentrations, slip velocities, and N_2 replacement by CO_2

- The particle size and gas composition dominate the ignition delay time via affecting the particle heating rate

- The N_2 replacement with CO_2 will not delay the ignition if the global gas temperature is constant

---

Nomenclature

Greek letters

β	aspect ratio (-)
ρ	gas density (kg/m^3)
ρ_p	particle density (kg/m^3)
ϕ_k	distribution coefficient (-)
Φ	mixture fraction (-)
ν	kinematic viscosity (m^2/s)
ν_{ji}	central wavelength (nm)
ϵ	area estimation error (-)
ϵ_p	particle emissivity (-)
ϵ_g	gas emissivity (-)
τ_i	ignition delay time of gas mixtures (ms)
σ	Stefan-Boltzmann constant (-)
Δt	time difference (ms)

Symbols

a	ellipse major axis (mm)
A	integral absorption (-)
A	pre-exponential factor for reaction kinetics (-)
A_e	area of a ellipse (mm^2)
A_p	particle area (mm^2)
A_{LU}	area of luminescence signals (mm^2)
b	ellipse minor axis (mm)
$B_{j,i}$	Einstein B coefficient (-)
c_p	heat capacity (constant pressure) (J/(kg·K))
c_v	heat capacity (constant volume) (J/(kg·K))
d_{90}	diameter of 90th percentile (µm)
d_p	circle-equivalent particle diameter (µm)
d_x	length of cubic cells (µm)
D_i	degeneracy factor (-)
f	focal length (mm)
$f_{B,i}(T)$	temperature dependent Boltzmann fraction (-)
Fo	Fourier number (-)

g_0	spectral line overlap integral (nm)
h	convection coefficient of heat transfer (-)
I^*_{LIF}	absorption-corrected LIF signals (a.u.)
L_d	distribution length (m)
m_p	instantaneous particle mass (kg)
$m_\mathrm{p,0}$	initial particle mass (kg)
Nu	Nusselt number (-)
Pr	Prandtl number (-)
q_c	the reaction heat release rate (J/s)
Re_p	particle Reynolds number (-)
t_ign	particle ignition delay time (ms)
$t_\mathrm{i,LU}$	first appearance of luminescence signal (ms)
t_vol	volatile combustion duration (ms)
$t_\mathrm{vol,end}$	end time of volatile combustion (ms)
t	residence time (ms)
T	optical transmission (-)
T_a	activation temperature (K)
T_ad	adiabatic flame temperature (K)
T_g	gas temperature (K)
$T_\mathrm{g,local}$	local gas temperature (K)
T_0	initial global gas temperature (K)
U_g	axial gas velocity (m/s)
U_p	axial particle velocity (m/s)
U_s	axial slip velocity (m/s)
Y_CH	CH mass fraction (-)
$Y_\mathrm{C_2H_2}$	C_2H_2 mass fraction (-)
$Y_\mathrm{CH_4}$	CH_4 mass fraction (-)
Y_F	fuel mass fraction (-)
$Y_\mathrm{F,0}$	initial fuel mass fraction (-)
Y_OH	OH mass fraction (-)

Abbreviations

CPD	chemical percolation devolatilization
COMS	complementary metal–oxide–semiconductor
DBI	diffuse-backlight illumination
DLSF	direct least squares fitting
DTF	drop tube furnace
EFR	entrained-flow reactor

FF	flat flame
FOV	field of view
FWHM	full width half maximum
hvb	high volatile bituminous
LFR	laminar flow reactor
LIF	laser-induced fluorescence
LU	luminescence
PDF	probability density function
PIV	particle image velocimetry
PTV	particle tracking velocimetry
ROI	region of interest
SAS	signal and structure
SNR	signal-to-noise ratio
SPC	single particle combustion
UV	ultraviolet

1. Introduction

On a global level, coal still is the most frequently used resource for electricity generation with a contribution of 36.4% in 2019 [1]. Among the several proposed technical solutions for reducing carbon dioxide (CO_2) emissions, oxy-fuel combustion is a promising technology for coal-fired power generation. This approach uses a mixture of pure oxygen (O_2) as an oxidizer combined with recirculated flue gas which results in a replacement of nitrogen (N_2) in air to CO_2 . Consequently, the flue gas mainly contains CO_2 and water (H_2O) which facilitates CO_2 capture and storage. However, the combustion processes are impacted significantly by introducing CO_2 as the major inert species due to its chemical and thermal pyhsical properties. This requires detailed understanding of the underlying multi-phase and physico-chemical sub-processes to enable the application of oxy-fuel combustion technology. Previous studies have been summarized in several reviews [2, 3, 4, 5, 6, 7] providing insights into fundamentals and industrial applications. In the present work, we focus on the fundamental processes of single particle combustion (SPC) in well-defined generic laminar flow conditions to gain a better understanding of ignition and volatile combustion. In the following, selected experimental and numerical studies on pulverized bituminous coal combustion in single-particle mode with high heating rates are briefly summarized.

1.1. A brief overview of experimental studies on single particle combustion

In the literature, drop tube furnaces (DTF) and flat flame burners (e.g. Hencken burners or laminar flow reactors) are suitable configurations to achieve high heating rates in the order of $\sim 10^5\,\mathrm{K/s}$. It has to be noted that experimental approaches for the definition of the ignition delay time and the volatile combustion duration are ambiguous. As differences in results may largely depend on the selected measurement techniques and their uncertainties, a comparison of different studies is only possible to a limited extent.

Timothy et al. performed temperature measurements of the sooting flame fueled with pulverized coals using two-color pyrometry in a laminar flow furnace [8, 9] with temperatures of 1250 K and 1700 K. Based on the temperature-time history and derived particle area, the devolatilization time is approximated by the time at which the particle temperature reaches its maximum. They found that the devolatilization time is shortened for increased oxygen concentrations. Bejarano et al. [10] and Khatami et al. [11, 12] conducted a series of experiments on single particle combustion in DTFs with a wall temperature of up to 1400 K using a three-color pyrometer. The volatile flame was investigated based on the typical two-peak temperature-time profiles associated to bituminous coal particles. They found that particle temperature increases with O_2 enrichment but decreases with increasing amounts of CO_2 in the atmosphere. Besides, soot formation is suppressed at high O_2 mole fractions as well as in CO_2 atmospheres [12]. However, the effect of particle size on temperatures of particle and gas flames is minor [10].

A restriction of the pyrometric methods is the lowest measurable temperature of typically around 1200 K. The particle temperature could also be possibly biased by the soot flame luminosity or the wall radiation due to the inherent nature of line-of-sight measurements. As recently developed alternatives, advanced optical imaging measurements provide further information for a thorough understanding of single particle combustion. Molina et al. investigated the ignition and volatile combustion of Pittsburgh high-volatile bituminous (hvb) coal in an entrained-flow reactor (EFR) with ambient gas temperatures T_g of $\sim 1250\,\mathrm{K}$ [13]. Time-averaged CH* signals recorded with an intensified CCD camera were used to derive the ignition delay time t_ign and the volatile combustion duration t_vol. It was concluded that O_2 enrichment significantly reduces t_ign. Further, the presence of CO_2 retards t_ign of single coal particles, whereas the volatile combustion duration t_vol is affected marginally. They reasoned that the increase in the volumetric heat capacity (the heat sink, ρc_p) of CO_2 leads to retarded gas-phase ignition. Shaddix et al. [14] conducted experiments in similar conditions by imaging the high-temperature sooting flame using an intensified CCD camera. Here, t_ign and t_vol are statistically evaluated by classifying flame images into different combustion stages according to soot shape and intensity. They concluded that increasing oxygen concentration accelerates particle ignition and volatile consumption for both N_2 and CO_2 atmospheres. Moreover, both t_{ign} and t_{vol} seemed to be greater when N_2 is replaced by CO_2. Khatami et al. [15] performed high-speed imaging in a DTF to detect ignition based on the soot flame luminosity. Their results showed that with increasing O_2 mole fraction, t_{ign} decreases significantly in a CO_2-enriched environment, while remaining almost unaffected in an N_2-enriched atmosphere.

However, soot particles are essentially produced in the middle-to-late stage of volatile combustion. Intermediate species, such as OH or CH radicals, have been seen as proper indicators for gas-phase flames and have been utilized to image the reaction zone using advanced laser diagnostics. Köser et al. applied high-speed imaging using planar laser induced fluorescence of the hydroxyl radical (OH-LIF) to visualize the single particle volatile flame in a laminar flat flame burner with gas temperatures of $\sim 1800\,\mathrm{K}$ [16, 17]. For the first time, the reaction zone near the particle-gas interface was visualized with high spatial and temporal resolution [16]. An increase of the O_2 concentration reduced the stand-off distance of volatile diffusion flames [17]. With an improved multi-parameter measurement approach for the simultaneous acquisition of particle size and flame topology [18], it was found that an increase of the particle diameter significantly impacts the volatile flame duration. Recently, a three-dimensional visualization using a laser scanning system has been applied to reconstruct the three-dimensional volatile flame topology [19, 20]. Gas-phase ignition was initially observed downstream of the particle, where high temperatures and fuel-lean mixtures exist. Increasing particle diameters resulted in larger stand-off distances of the main reaction zone and hence a larger flame size.

1.2. A brief overview of numerical studies on single particle combustion

In the literature related to numerical studies on pulverized coal combustion, it has been found that the devolatilization process in coal combustion is one of the most challenging parts of direct numerical simulation and requires detailed models to simulate the physico-chemical behavior correctly. A well established model for the devolatilization process is the chemical percolation devolatilization (CPD) model which describes the effect of the molecular structure of coal on the devolatilization process in detail [21]. This devolatilization model couples finite rate chemistry with a detailed mechanism for gas phase reactions.

Recently, numerical simulations for single [22] and multiple interacting particles [23] have been performed employing the CPD model [21] in a fully coupled Eulerian-Lagrangian point-particle numerical framework to investigate a number of aspects including the sensitivity of ignition delay time to temperature, gas composition, particle size, and particle number density. The results have been validated against a series of experiments in a laminar entrained flow reactor [24]. A double peak profile in the time evolution of the volatile release rate has been observed, which strongly affects the ignition delay time. It has been shown that ignition can occur on the onset of the first or the second peak of the devolatilization rate in different initial gas temperatures and particle sizes. It has also been shown that particle Reynolds number and slip velocity between gas and particles have an impact on ignition delay time and the volatile flame which cannot be neglected. The formation of wake flames due to high slip velocities becomes increasingly relevant for flame interactions with multiple interacting particles [25] and turbulent coal combustion [26]. Switching from conventional to oxy-fuel combustion can affect the combustion behavior due to the different thermal and chemical properties of CO_2 compared with N_2. For instance, several numerical [27, 28, 29] studies indicate, that replacing N_2 by CO_2 increases the ignition delay time and the duration of devolatilization.

1.3. Objective of the present work

The present work's aim is the investigation of relevant physico-chemical effects on pulverized coal combustion on a single particle scale. For this purpose, the ignition and volatile combustion of single bituminous coal particles are examined employing both experimental and numerical methods under well-defined laminar flow conditions. Because consecutively performed measurements of different quantities can only be combined using statistics with the need for an evaluation of the repeatability of the investigated process, multi-parameter measurements are conducted in this study such that relevant quantities are measured simultaneously providing a comprehensive data set with high spatial and temporal resolution. The volatile flame is visualized by using OH-LIF to detect the ignition and using luminescence imaging to determine the temporal end of volatile combustion. Additionally, high-speed backlight illumination addresses particle size, shape and velocity. With well defined boundary conditions, experimental results are analyzed showing the importance of particle size and ambient gas composition. Detailed numerical simulations are conducted and validated against experiments with respect to the ignition delay time and volatile combustion duration. Beyond that, simulations provide further insights into particle temperatures, local gas temperatures, volatile release rates, and fuel mass fractions. Combining experimental and numerical expertise, effects of particle size, oxygen concentration, slip velocity and the introduction of a CO_2 atmosphere are discussed to understand the ignition and volatile combustion in the present configuration.

The structure of this work is as follows: in Section 2, experimental configurations and data processing procedures are introduced. Subsequently, numerical models and methods are outlined in Section 3. Experimental results are presented and briefly discussed in Section 4. To better understand the observed phenomena, numerical results are included in the discussion in Section 6. Validations of several cases are presented considering realistic inlet boundary conditions. On this basis, physico-chemical processes during the devolatilization and the importance of different parameters are discussed. Finally, the main outcome of the study is summarized.

2. Experimental methodology

2.1. Laminar flow reactor

Experiments were performed in a laminar flow reactor (LFR) with well-defined boundary conditions at the Technical University of Darmstadt. The configuration was introduced in previous works [17, 18] and is illustrated in Fig. 1(a). The LFR consisted of a 80×80 mm ceramic honeycomb structure, a fused silica enclosure for optical access and a particle seeding unit with a central injection tube with an inner diameter of 0.8 mm. A premixed laminar flat flame (FF) was stabilized above the burner surface. Both the flat flame and particle carrier gas (Jet) were operated with identical inlet gas mixtures. In this study, four N_2 atmospheres (denoted as AIR) with different inlet mixtures of $CH_4/O_2/N_2$ and three CO_2 atmospheres (denoted as OXY) with varying inlet mixtures of $CH_4/O_2/CO_2$ were investigated. Table 2 lists the oxygen concentration in the post-flame environment, flow rates, equivalence ratio Φ, and adiabatic flame temperatures T_{ab} for each condition. Based on the 1D calculation, the produced water vapor concentration was ~ 14 vol% and ~ 16 vol% in AIR and OXY conditions, respectively. Cai et al. [30] reported that particle ignition is not affected by a H_2O mole fraction up to 20 vol% under comparable conditions. Hence, the influences of water vapor concentrations are considered marginal in this study.

The point of heat-up of individual fuel particles was precisely defined by the particle crossing the flame front of the premixed CH_4 flame. The flame position, as illustrated in Fig. 1(c), was previously characterized for each condition with a separate OH-LIF measurement. It should be noted that the flat flame revealed a slightly asymmetry which was mainly caused by the inlet flow and the heat loss to the ceramic surface. For all cases but OXY20, which is shown in Fig. 1(c), this imperfection was hardly noticeable as relatively symmetric flame structures were present. When a bituminous coal particle entered the oxygen-enriched high-temperature environment, it experienced a heating rate in the order of $\sim 10^5$ K/s [18]. This induced a rapid water evaporation followed by the successive release of volatiles, gas-phase ignition, volatile combustion and char combustion. By modulating the wheel rotation speed of the seeding unit, individual particles were sparsely seeded to guarantee single particle combustion. To give an impression of particle combustion in the LFR, luminous streaks of burning particles are shown in the color image of Fig. 1(b).

Colombian high-volatile bituminous coal particles with two different sievings were used in this study. The samples, denoted as A and B, were sieved to a range of 90 to 125 µm and 160 to 200 µm, respectively. The proximate analysis of the coal composition was 3.5%m moisture(an), 36.9%m volatiles(wf), 54.4%m Cfix(wf) and 8.7%m ash(wf).

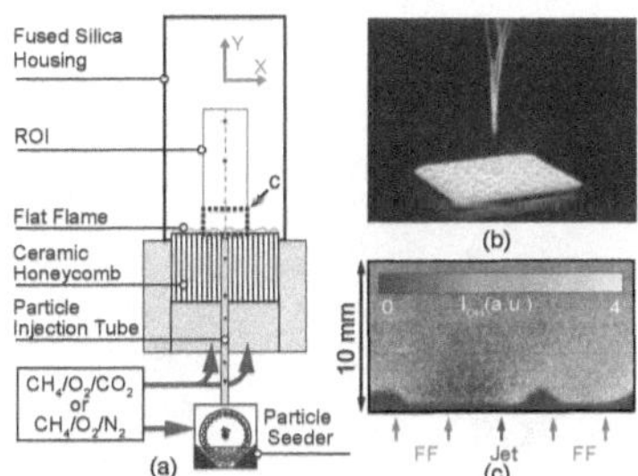

Figure 1: (a) Sketch of the Darmstadt laminar flow reactor. (b) Luminosity photograph of burning particles and the flat flame. (c) 2D OH-LIF visualization of the flat flame structure for OXY20 in the region highlighted in (a) by dashed lines.

Table 2: Inlet flow rates defined for different AIR and OXY conditions.

	O$_2$ content	Φ (-)	**Flat flame** (ln/h)			**Particle jet** (mln/min)			T$_{ad}$ (K)
			CH$_4$	N$_2$	O$_2$	CH$_4$	N$_2$	O$_2$	
AIR10	10 vol%	0.58	312	3203	1085	1.12	11.49	3.89	1847
AIR20	20 vol%	0.40	312	2744	1543	1.12	9.85	5.54	1839
AIR30	30 vol%	0.31	311	2286	2003	1.12	8.21	7.18	1829
AIR40	40 vol%	0.25	313	1822	2286	1.12	6.54	8.84	1829
			CH$_4$	CO$_2$	O$_2$	CH$_4$	CO$_2$	O$_2$	
OXY20	20 vol%	0.47	307	1874	1315	1.44	8.86	6.19	1835
OXY30	30 vol%	0.37	306	1553	1672	1.43	7.25	7.81	1881
OXY40	40 vol%	0.29	305	1310	2094	1.36	5.83	9.31	1825

2.2. Measurements of boundary conditions

2.2.1. Flow field

The laminar flow field was measured using high-speed particle image velocimetry (PIV) as shown in Fig. 2. The carrier gas was seeded with Al_2O_3 particles (MR-52, Martoxid, $d_{90} = 3 \sim 6\,\mu m$) through the particle injection tube. The seeding particles were illuminated by a frequency-doubled Nd:YAG laser (Innoslab, Edgewave) with 2 mJ per pulse at 10 kHz. The laser beam was formed to a sheet with a height of 25 mm and focused at the burner center line to a thickness of $\sim 100\,\mu m$. The Mie scattering signals were collected by a CMOS camera (Fastcam SA-X2, Photron) equipped with a macro lens (Sigma, $f = 180\,mm$, $f/16$). A band-pass filter ($532 \pm 5\,nm$) was employed to suppress flame luminosity. The FOV was $20 \times 10\,mm^2$ (height $\times$ width) with a pixel resolution of $20\,\mu m$.

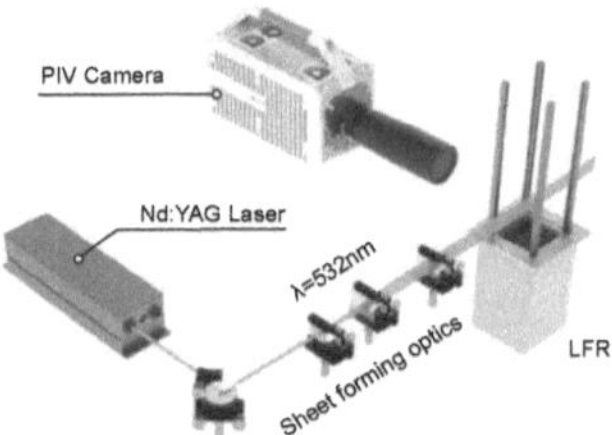

Figure 2: Gas velocity field characterization: experimental setup for the high-speed PIV measurements.

The laminar flow field was evaluated using a combined PIV-PTV approach (Davis 10, LaVision) with temporally equidistant images of the seeded flow. The particle vectors were preliminarily evaluated using a PIV algorithm including a multi-pass cross-correlation calculation with gradually decreasing interrogation window sizes. For the subsequent PTV processing, a correlation window size of 16×16 pixel with a vector tolerance of 2 pixels relative to the PIV calculation was applied. The measurements took place at various height positions, such that the region of up to 50 mm above the burner was covered.

2.2.2. Temperature field

Within this study, quantitative OH-PLIF combined with absorption spectroscopy was used to derive a local temperature field. This was possible for lean flames with $\Phi < 0.9$ due to the unambiguous correlation between OH number density and temperature in chemical equilibrium [31]. The correlations have been computed using

0 D reactor simulations for the given atmospheres with varying enthalpy using CAN-TERA [32]. This approach circumvented the usual quantification problems for LIF techniques (e.g. unknown chemical surrounding, detecion efficiency). The absorption of laser energy due to resonant energy transfer to the OH molecule was derived by measuring an energy reference before and after the probe volume. The energy ratio for on-resonant measurements was normalized with the off-resonant energy ratio to eliminate influences from non-resonant losses, e.g. on windows and lenses. The integral absorption was calibrated using spectroscopic data from LIFBASE [33] using Eq. 1,

$$\int [OH](x,y)\mathrm{d}r = \frac{A(y)}{g_0 \frac{h\nu_{ji}}{c} B_{ji} D_i f_{B,i}(T)},\tag{1}$$

where A denotes the measured integral absorption, g_0 the spectral line overlap integral, ν_{ji} the central wavelength, B_{ji} the Einstein B coefficient, D_i the degeneracy factor and $f_{B,i}(T)$ the temperature dependent Boltzmann fraction of the chosen absorption line.

The integral absorption was then distributed along the beam-wise direction x using the absorption corrected LIF signal $I^*_{LIF}(x,y)$ itself, see Eq. 2.[R2]

$$[OH](x,y) = \frac{I^*_{\mathrm{LIF}}(x,y)}{\int I^*_{\mathrm{LIF}}(x,y)\mathrm{d}x} \int [OH](x,y)\mathrm{d}x.\tag{2}$$

This approach cancels all influences on the LIF signal that are independent of the coordinate x for any given height y. In particular, as long as the fluorescence quantum yield is constant along the beam-path, influences from quenching are canceled. This method yields a very high sensitivity and precision, as the equilibrium OH number density doubles approximately every $100\,\mathrm{K}$ under these conditions. However, temperature measurements are limited to areas with significant OH content, i.e. $> 1400\,\mathrm{K}$. A proper selection of the absorption line is crucial for this approach. This is especially true for atmospheric conditions, where the absorption linewidth is in the same order of magnitude as the laser linewidth. Additionally, the Boltzmann fraction should not vary greatly for the temperature range expected in the investigation. For these reasons, the $Q_1(6.5)$ transition in the OH $X^2\Pi \rightarrow A^2\Sigma$ ($\nu'' = 0 \rightarrow \nu' = 1$) at $283.01\,\mathrm{nm}$ system was chosen. The Boltzmann fraction for this line varies $\approx 10\%$ in the temperature range between $1600 - 1850\,\mathrm{K}$.

The optical setup is schematically depicted in Fig. 3. To excite the OH molecule, a high-speed diode pumped solid state Nd:YAG laser (Innoslab, Edgewave) was used to pump a dye laser (Allegro, Sirah) operated with Rhodamin 6G dissolved in ethanol. The fundamental wavelength and linewidth was measured using a wavemeter

(WSU-30, HighFinesse) which was calibrated using a temperature stabilized single mode HeNe diode laser. The laser system was operated at 500 Hz repetition rate. OH fluorescence was detected using a high-speed CMOS camera (HSS6, LaVision) coupled with a two-stage image intensifier (HS-IRO, Lavision). A 100 mm lens (2178, Cerco, $f/2.8$) was employed for high UV-transmission. A narrow-band band-pass filter (T $\geq$ 90% @ $310-320$ nm) was used to suppress ambient light as well as Rayleigh scattering. The filter transmits only the (1,1) vibrational band to minimize signal re-absorption [34]. For the absorption measurement, a fraction of the laser beam before and after the probe volume was extracted and guided to a reference cell filled with a diluted Rhodamin 6G ethanol solution. Beam steering effects were minimized by introducing a 4f-imaging lens in the post-burner beam path. Fluorescence was detected using a second high-speed camera (Fastcam SA-X2, Photron) equipped with a 180 mm lens (Sigma, $f/5.6$).

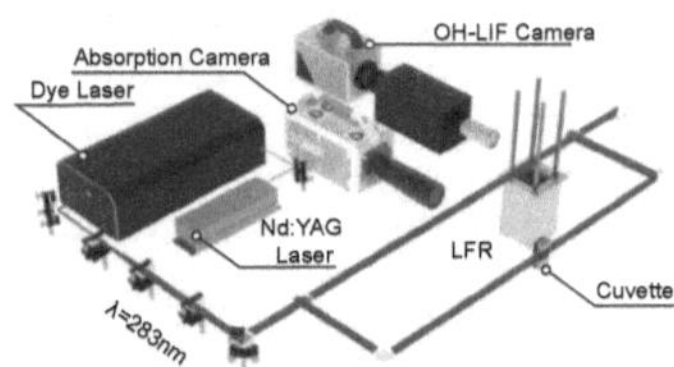

Figure 3: Gas temperature field characterization: experimental setup for the combined OH-PLIF/absorption spectroscopy measurements.

2.3. Multi-parameter measurements of single particle combustion

Simultaneous 10-kHz multi-parameter optical measurements were performed including planar OH-LIF imaging, diffuse-backlight illumination (DBI) and luminescence (LU) imaging. Figure 4 illustrates the experimental setup including the arrangement of the three diagnostic systems. For the reaction zone visualization using OH-LIF, the dye laser system introduced for the temperature measurements was employed at a repetition rate of 10 kHz. The wavelength was tuned to 283.01 nm to excite the $Q_1(6.5)$ line of the A-X(1-0) transition system of the OH radical. The effective pulse energy in the probe volume was approximately 0.3 mJ. The laser beam was formed into a sheet with a thickness of 0.8 mm (FWHM) at the burner center line. The aim of this relatively thick laser sheet was to reduce the probability of out-of-plane loss particles and to allow the signal detection in the early stage of volatile flames. The emission signals were collected by the aforementioned high-speed intensified camera (HSS6 + HS-IRO, LaVision) equipped with a UV-achromatic camera lens

(Halle, $f = 150\,\text{mm}$, $f/2.5$). To suppress broadband chemiluminescence and thermal radiation from volatile flames, a band-pass filter (T $\geq 40\%$ @ 305 - 340 nm) was used and the intensifier gate was shortened to 100 ns. The field of view (FOV) for OH-LIF imaging was $19 \times 19\,\text{mm}^2$ with a projected pixel size of 25 µm.

For particle shape and size characterization, a DBI system including a CMOS camera (HSS6, LaVision) and a high-power LED (IPS, ILA) was inclined by 17° to the OH-LIF detection system. The LED was operated at 10 kHz with a peak wavelength of 525 nm and a pulse duration of 1 µs. The CMOS camera equipped with a long-distance microscope (SK2, Infinity) and a band-pass filter (525 ± 25 nm) detected the particle shadow images with high spatial resolution. With a projected pixel size of 10 µm, the FOV of $13 \times 5\,\text{mm}^2$ (height × width) covered the region from the LFR surface to 13 mm along the y-axis.

To track the temporal evolution of burning particles, another CMOS camera (Fastcam SA-X2, Photron) combined with a macro lens (Sigma, $f = 180\,\text{mm}$, $f/5.6$) was employed to image the luminous flame. A band-pass filter (380 - 492 nm, Semrock) was used and the recorded intensity was dominated by the thermal radiation of soot, tar and particle surface [18]. The FOV of $65 \times 10\,\text{mm}$ (height × width) allowed a particle tracking throughout the entire volatile combustion stage. Due to the large field of view, the projected pixel resolution was restricted to 60 µm but was still sufficient to resolve volatile flames with a size of a few millimeters.

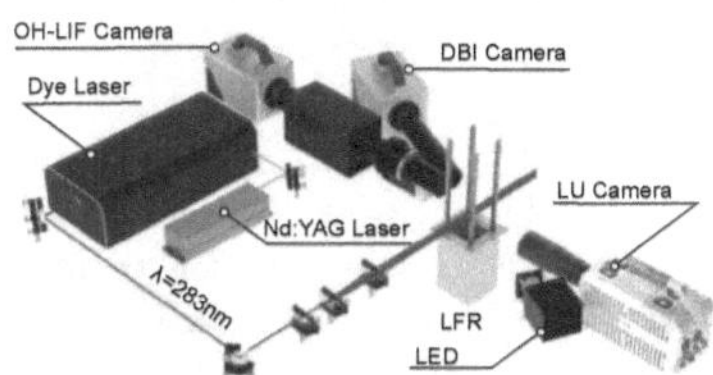

Figure 4: Experimental setup of simultaneous multi-parameter measurements using OH-PLIF, LU and DBI.

2.4. Multi-parameter data processing

The utilization of multi-parameter measurements is essential to obtain case-specific information about ignition and volatile combustion. This is important as the measured quantities of interest are directly linked to each other, enabling a thorough statistical analysis with reduced complexity. Correct data interpretation and better understanding of the underlying sub-processes rely on appropriate data processing strategies. Several crucial parameters relevant to single particle combustion are derived and subsequently, corresponding processing procedures are described in detail.

As one of the key quantities, the ignition delay time t_{ign} is determined by temporally and spatially tracking the OH-LIF signal. Instantaneous laser profiles are obtained from the background OH-LIF intensity associated with the post-flame region, where a homogeneous distribution of OH-radicals from water dissociation is assumed. The OH-LIF images are shot-for-shot corrected in terms of the pulse energy fluctuation and the sheet inhomogeneity. The signal-to-noise ratio (SNR) is estimated to be approximately 12 and the maximum signal-to-background ratio is around 10. The signals are normalized based on the background intensity and further denoised by using a 5×5 Gaussian filter. The particle centroids are extracted from simultaneous DBI recordings and used as position references for particle tracking. Only the particles with a sufficient distance of $\geq 4\,\text{mm}$ to the nearest particle were included in the data analysis which guaranteed single-particle combustion mode. As shown in Fig. 5 (a,c,d), a $2 \times 2\,\text{mm}^2$ region of interest (ROI, dashed line) is defined centered around the particle centroid (red cross) in the pre-processed OH-LIF images. By evaluating the background intensity, an intensity threshold of 1.2 is applied for binarization in Fig. 5 (b,d,e). Within the ROIs, a signal and structure (SAS) analysis was performed. The ignition time is determined to when (1) the mean OH-LIF intensity increases over the threshold and (2) the connected area of the largest binary structure exceeds $1 \times 1\,\text{mm}^2$. The latter criterion is defined based on an estimation of the spatial resolution restricted to the laser sheet thickness. Figure 5 illustratively shows three instants at $t_{\text{ign}} - 0.5\,\text{ms}$, t_{ign} and $t_{\text{ign}} + 0.5\,\text{ms}$ for a single-particle ignition event. The heating start time is set to $t = 0$ and is defined at the instant at which a particle crosses the flame front of the premixed CH_4 flame.

The volatile flame duration t_{vol} is determined by combining t_{ign} from OH-LIF measurements and the end time of volatile combustion $t_{\text{vol,end}}$ from luminescence measurements. The particle positions extracted from DBI images are used to initialize the temporal tracking of LU signals. The area of LU signals A_{LU} is evaluated by defining an intensity threshold which is placed marginally (i.e. $\sim 10\%$) above the background level. Figure 6(a) shows the temporal variation of the LU signal area of a single particle flame. Four individual instants are highlighted and visualized in Fig. 6(b-e). The color scale has been adapted to match the low intensity level. A few milliseconds after the onset of ignition, the size of the gas flame increases and then decreases to a local minimum. This implies a transition from the formation of a diffusion flame to the completed fuel consumption via gas-phase oxidation [18]. In

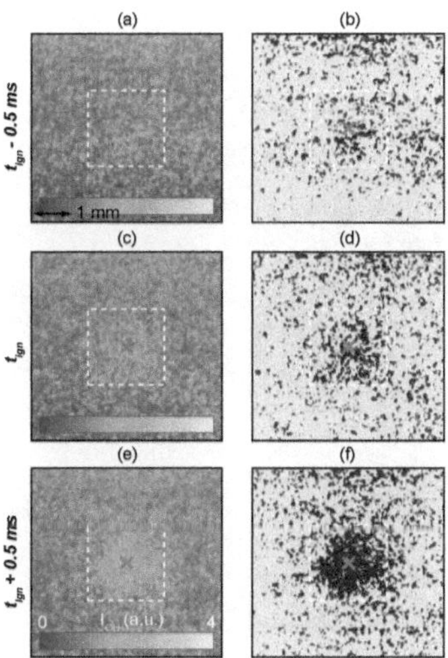

Figure 5: An example of ignition detection using temporal OH-LIF signals at (a, b) 0.5 ms before ignition, (c, d) ignition and (e, f) 0.5 ms after ignition. Left column: pre-processed OH-LIF images superposed with particle centroids (red crosses) from simultaneous DBI images and a $2 \times 2\,\mathrm{mm}^2$ region of interest (dashed lines). Right column: corresponding binarized OH-LIF images.

the subsequent instants, no significant increase in A_{LU} is noticeable since luminous signals are dominated by particle surface reactions. However, the luminosity intensity increases again (not shown) due to particle surface reactions. Hence, the local minimum area is used for an appropriate determination of the end time of volatile combustion $t_{\mathrm{vol,end}}$. In a few cases, the luminosity intensity disappears at the end of volatile combustion and appears after char combustion begins. The $t_{\mathrm{vol,end}}$ is then determined at the time point of the first disappearance of luminosity signals. Based on this, the pre-ignition, volatile combustion and char combustion stages are temporally classified for each single particle, as shown in Fig. 6. Consequently, the entire volatile combustion duration is defined as $t_{\mathrm{vol}} = t_{\mathrm{vol,end}} - t_{\mathrm{ign}}$.

The 2D projected particle size and shape are characterized with an in-situ DBI measurement. Figure 7 shows three representative particles with shape parameters given in the table. The boundary detection based on the steep gradient is performed using an adaptive thresholding method proposed in previous work [35, 36]. The

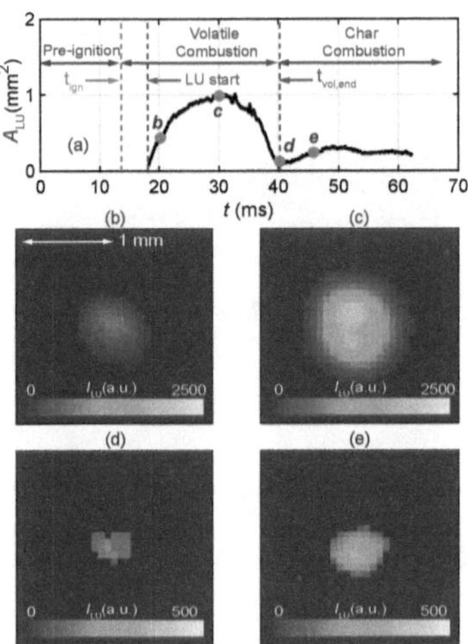

Figure 6: (a) Temporal variation of luminescence signal area A_{LU} of a single particle. (b-e) Four individual 2D visualization of luminescence intensity at successive time steps indicated in (a).

circle-equivalent particle diameter d_{p} is evaluated based on the enclosed particle area A_{p}. The particle centroids are evaluated and further utilized for particle tracking in simultaneous OH-LIF and LU measurements as discussed before. The particle shape is approximated by an ellipse employing direct least squares fitting (DLSF). Using the major and minor axis, denoted as $2a$ and $2b$ respectively, the aspect ratio is defined as $\beta = a/b$. The error of the area estimation by using the ellipse fitting algorithm is indicated by $\epsilon = (|A_{\mathrm{p}} - A_{\mathrm{e}}|)/A_{\mathrm{p}}$, where A_{e} is the area of the ellipse. Comparing the three particles in Figure 7, particle P_1 and P_2 have a similar shape (β), whereas particle P_2 and P_3 have similar size (d_{p}); this is in accordance with observations. Hence, it can be concluded that the selected parameters are appropriate for characterization of particle shape and size in this study.

3. Numerical framework and modeling

In the present work, pulverized coal combustion is modeled in an Eulerian-Lagrangian framework, using an Eulerian formulation for the gas phase and a Lagrangian formulation with point particle approximation for the solid particles, which

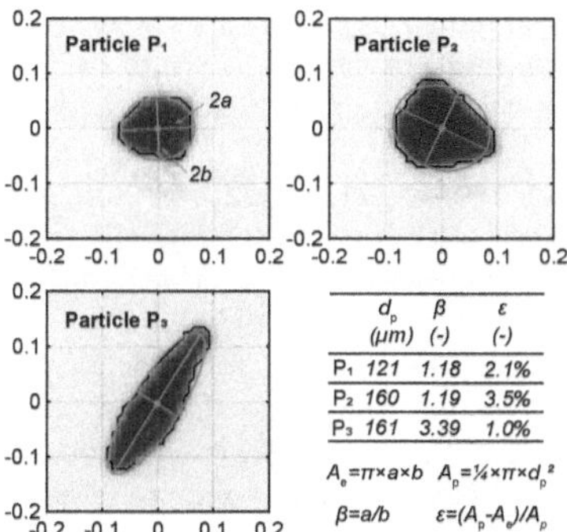

	d_p (μm)	β (-)	ϵ (-)
P₁	121	1.18	2.1%
P₂	160	1.19	3.5%
P₃	161	3.39	1.0%

$A_e = \pi \times a \times b \quad A_p = \frac{1}{4} \times \pi \times d_p^2$

$\beta = a/b \quad \epsilon = (A_p - A_e)/A_p$

Figure 7: DBI images of there representative particles and definitions of relevant parameters. d_p: circle-equivalent diameter, β: aspect ratio of the fitted ellipse, A_p: particle area, A_e: area of the fitted ellipse, ϵ: error of the particle area approximated by the ellipse fitting.

are fully coupled using a two-way coupling approach. The models were described in detail by Farazi et al. [22, 23]. The governing equations in the gas phase are similar to those applied by Sirignano [37], Attili et al. [38], and Bai et al. [39] with a low Mach number assumption of the Navier-Stokes equations and obeying the ideal gas law for the gas mixture. Chemical reactions in the gas phase are modeled with a finite rate chemistry model using the ITV oxyflame mechanism for methane with 68 species and 906 reactions [40]. Gas-phase radiation is included using the optically-thin approximation. The Eulerian governing equations are solved by using a semi-implicit finite difference code with second-order accuracy in space and time [41] and the Poisson equation is solved applying the multi-grid solver of HYPRE. The Crank-Nicolson method is applied for time advancement along with an iterative predictor-corrector scheme [42].

Coal particles are modeled by the Lagrangian framework, and the equations for determining mass, trajectory, velocity, and temperature of the particles are similar to those applied by Farazi et al. [23]. To describe the devolatilization process, the chemical percolation devolatilization (CPD) model is applied. This model determines the devolatilization rate and composition of tar and light gases based on bond breaking in the molecular structure of reference coals as a function of time according to the study by Grant et al. [21]. In the present work, the light gases consist of CH_4, CO_2, CO, H_2O, and other gases. Other gases are assumed to be C_2H_2, similar to the assumption by Jimenez and Gonzalo-Tirado [29]. The rate of tar release, which is also computed by the CPD model, is assumed to be only for C_2H_2. This assumption is also used by Goshayeshi and Sutherland [43] and Tufano et al. [28]. To assess the effect of this assumption on the ignition delay time for the investigated coal and setup, other species like C_6H_6 were used as tar and only a marginal 3% difference in ignition delay time has been observed [22]. Although the ignition delay time of C_2H_2 in a purely gas phase setting is much lower than that of C_6H_6, the marginal difference in t_{ign} for the particle setting shows that ignition time is more dominated by the characteristic time required for particle heating. Uncertainties regarding the choice of tar species are taken into account for calculating the ignition delay time.

The coupling between the gas phase and the solid phase has been done through the source terms appearing in the conservation equations by means of a distribution coefficient ϕ_k for each particle, which is computed by a Gaussian function with characteristic width L_d. In this approach, similar to the one used by Farazi et al. [22, 23], the distribution length L_d is set to $2d_p$ where d_p corresponds to the particle diameter. It has been shown that extending L_d from d_p to $5d_p$ (corresponding to a volume of $125dx^3$) changes the ignition delay time by less than 10%. With the larger values of L_d, the source term is distributed far beyond the reaction zone around each particle, which leads to a significant influence on the ignition delay time [22]. In the Eulerian-Lagrangian approach, the particle equations are derived according to the film model, assuming a uniform gas field around the particle. The domain size is discretized using a uniform grid with cubic cells of length d_x, which is equal to the particle diameter. A convergence study by Farazi et al. [22] showed that for $d_x < d_p$, the ignition delay time does not change, and for $d_x > d_p$, the homogeneous combustion cannot be captured around single particles and would lead to a very dilute mixture which could prevent ignition. As a result, $d_x = d_p$ is used for all simulations. Also, since using the gas phase quantities from grid cells with the same size as the particle might not be consistent with the film model assumption, a filter is applied to provide a smoother field in the gas phase to evaluate the state of the gas surrounding the particle consistently with the film model. The sensitivity studies by Farazi et al. [22] showed that the filter length does not have a significant effect on the results. The accuracy of the models and the methods has been validated against experimentally measured ignition delay times by Liu et al. [24].

For the conducted numerical simulations, an inlet-outlet configuration based on the experimental setup of the flat flame burner is considered. Particles are injected into a hot gas-mixture stream with thermodynamical conditions obtained from the fully burned composition and temperature behind an unstretched premixed laminar methane flame. The initial velocity of the gas stream and particle, temperature, size, and composition are prescribed using experimental data and evolve due to the particle-gas interaction. In the present work, the flat flame region is neglected because it is assumed to have a negligible effect on the particle temperature and ignition because of the small thickness of the flame and, as a result, the residence time of the particle inside the flat flame is small compared to the ignition time.

4. Experimental results

4.1. Particle size and shape

Figure 8 shows the probability density functions (PDF) of the particle diameter d_p (a), the aspect ratio β (b) and the error of the DLSF approximation ϵ (c) for investigated samples A and B. Approximately 1000 particles from different atmospheres are included for each distribution. To avoid particle swelling and beam steering, only the first appearance of particles after crossing the flame front, which stabilizes approximately at $y = 2\,\mathrm{mm}$, is considered in this analysis. Figure 8(a) shows that both samples deviate from the sieving specifications in the sense that the size distributions are broader and shifted towards larger diameters as coal particles are not perfectly spherical and therefore sometimes pass the sieve although having a larger circle-equivalent diameter than the sieving mesh width. The mean particle diameters (dashed line) are $120\,\mathrm{\mu m}$ and $205\,\mathrm{\mu m}$ for A and B respectively. For β and ϵ, no significant discrepancy between differently sieved particles is noticeable. The highest probability of β is observed for a value of 1.5 and most particles have an aspect ratio of $\beta \leq 3$. The DLSF particle approximation stays mostly within an acceptable error margin of 5%.

4.2. Gas Temperatures

The mean gas temperature and standard deviation (bar) along the burner center line are show in Fig. 9 for (a) AIR and (b) OXY conditions. The results are computed for discrete positions along y-axis with a equidistant subdivision with a width of $5\,\mathrm{mm}$. The initial temperatures at $y = 2.5\,\mathrm{mm}$ are in good accordance with the adiabatic flame temperatures evaluated by a 1D simulation. Crossing the main reaction zone of the premixed CH_4 flame, a noticeable decrease in the slope of T_g at $y = 7.5\,\mathrm{mm}$ is observed for all conditions. Downstream of this, T_g decreases linearly with increasing y with a slop of $\sim 2\,\mathrm{K/mm}$. Since the differences in the flow rate of CH_4 are minor, as shown in Table 2, thermal energy produced by premixed combustion is expected to be similar for different conditions. Due to the higher heat capacity of CO_2, the entire flow rates in OXY conditions are reduced in order to increase the gas temperature. As one can see in Fig. 9, the temperature profiles along the center line are similar for all investigated atmospheres, which enables reasonable comparisons of single particle combustion.

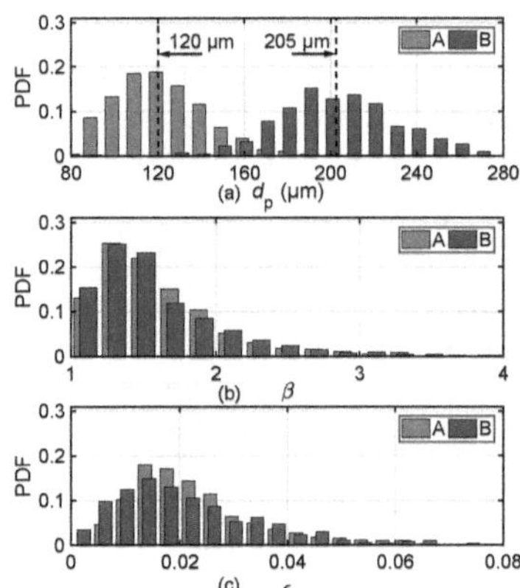

Figure 8: PDF distributions of the particle diameter d_p, the aspect ratio β and the error of the DLSF approximation ϵ.

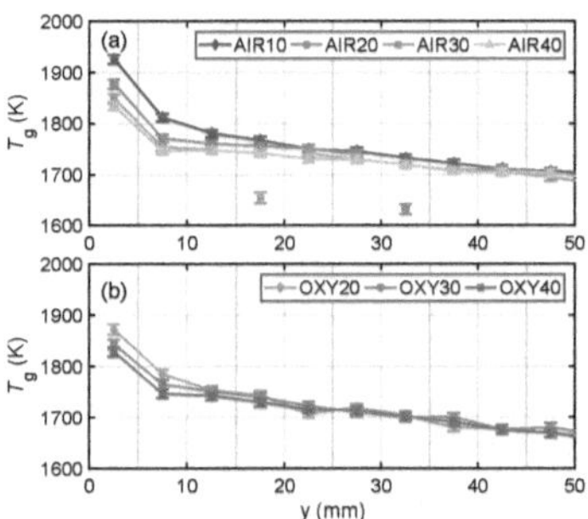

Figure 9: Gas temperature T_g over the height y along the burner center line in (a) AIR and (b) OXY conditions. (Outliers at $y = 17.5\,\mathrm{mm}$ in AIR30 and at $y = 32.5\,\mathrm{mm}$ in AIR20 due to failures/errors of the absorption measurements.)

4.3. Gas and particle dynamics

Figure 10 shows the axial gas velocity U_g along the vertical coordinate of the burner for (a) AIR and (b) OXY atmospheres. The bars indicate one standard deviation at each y position. A steep gas acceleration is observed within the first $2\,\mathrm{mm}$, which spatially correlates to the pre-heat zone and main reaction zone of the premixed

CH$_4$ flame. As a result, the gas accelerates due to the steep temperature gradient and thermal expansion across the flame. Since the seeding particles are not uniform in size and hence show a deviating following behavior within the gas flow, a relatively high standard deviation of the measured velocity is observed within the flame. More importantly, high gas velocity gradients exist in the flame region which mainly result in a larger standard deviation.. From approximately $y = 3$ mm, the velocity profile starts to stabilize and then remains relatively constant. Comparing different atmospheres, the stabilized gas velocity of AIR conditions is approximately 1.6 m/s and greater than the velocity of 1.3 m/s for OXY conditions. This is attributed to the lower overall flow rates of OXY compared to AIR conditions, as listed in Table 2, in order to obtain similar gas temperature profiles. AIR conditions with different O$_2$ concentrations reveal similar velocity profiles; the same is observed for all OXY conditions as well. A slight discrepancy is noticed for OXY20 at $y = 2$ mm. This is a result of the unevenness of the flat flame structure which is slightly lifted in this condition, as implied in Fig. 1(c). Therefore, seeding particle trajectories might be disturbed by the semi-spherical flame shape close to the particle jet.

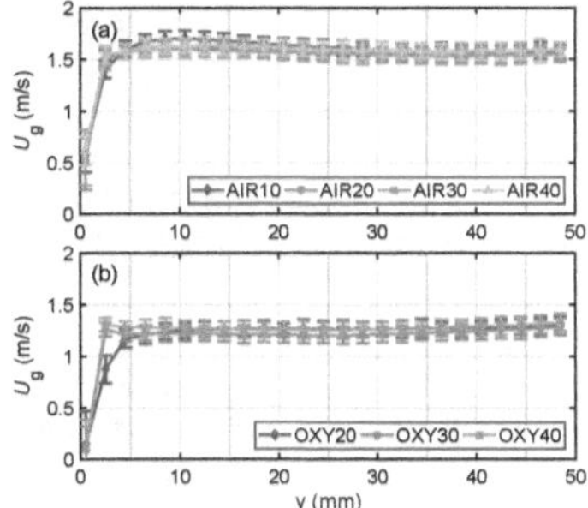

Figure 10: Axial gas velocity U_g over the height along the burner center line y for (a) AIR and (b) OXY conditions.

The particle velocity is evaluated using the temporal evolution of the particle position obtained from DBI measurements. The axial velocity component U_p is calculated with the first derivative of the y component of the particle position using its five-point stencil. Figure 11 shows the statistical results of both particle samples under AIR (a) and OXY (b) conditions. By subdividing the y-axis into segments with a width of 1 mm each, it is possible to compute mean values of samples included in the respective segment with standard deviations indicated by bars at each position along y. As Figure 11 illustrates, the smaller particles of sample A show a higher initial velocity after crossing the flame front due to their lower mass inertia compared to the larger particles of sample B. Afterwards, the velocity slope of both particle sizes is similar, however, smaller particle have a higher acceleration considering the shorter residence time compared with larger particles. This points to the fact that the particle velocity history is essentially dominated by the steep gas velocity gradient within the flat flame, while the aerodynamic drag downstream is of secondary importance.

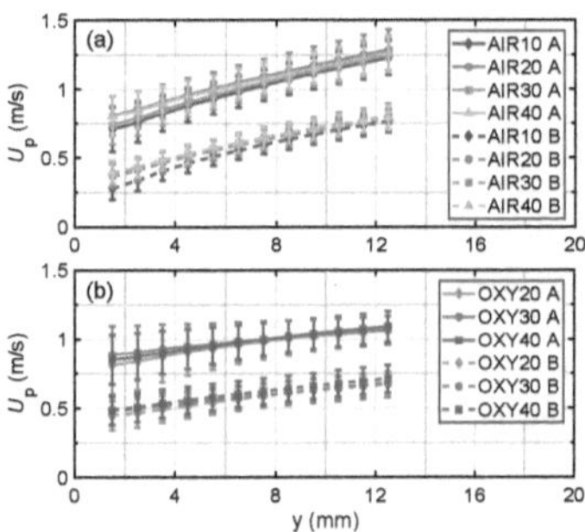

Figure 11: Particle velocity U_p over the height y in (a) AIR and (b) OXY conditions for differently sized particles A (mean $d_\mathrm{p} = 120\,\mu\mathrm{m}$) and B (mean $d_\mathrm{p} = 205\,\mu\mathrm{m}$).

With knowledge of the mean gas and particle velocities, the mean slip velocity U_s can be calculated and is depicted in Fig. 12 for all atmospheres and both size samples. U_s first increases over the CH_4 flame and then decreases with the stabilization of the gas velocity after $y = 3\,\mathrm{mm}$. The particle Reynolds number can be estimated using the slip velocity and the particle diameter to $Re_\mathrm{p} = U_\mathrm{s}d_\mathrm{p}/\nu$, where ν denotes the kinematic viscosity of the gas mixture. Based on the gas composition and the adiabatic flame temperature, ν is estimated to be $\sim 3.3 \times 10^{-4}\,\mathrm{mm}^2/\mathrm{s}$ for AIR conditions and $\sim 2.7 \times 10^{-4}\,\mathrm{mm}^2/\mathrm{s}$ for OXY conditions. Using the maximum slip velocity and mean particle diameter, the particle Reynolds number of small particles is estimated to be 0.29 in AIR and 0.17 in OXY conditions; the particle Reynolds number of large particles is 0.73 in AIR and 0.60 in OXY conditions.

4.4. Comparison of methods for ignition detection

Experimental results of ignition delay times always depend on the uncertainties of both the selected measurement techniques and data processing methods. Köser et

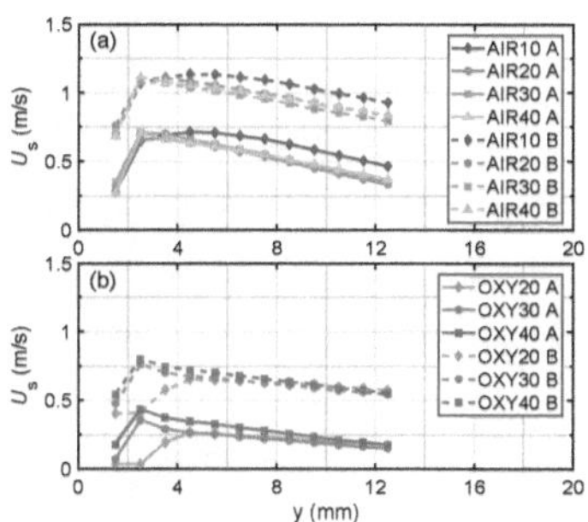

Figure 12: Slip velocity U_s over height y in (a) AIR and (b) OXY conditions for differently sized particles A (mean $d_p = 120\,\mu m$) and B (mean $d_p = 205\,\mu m$).

al. compared simultaneous OH-LIF and luminescence images at the onset of particle ignition and concluded that the luminescence technique overestimated the ignition delay time due to a strong dependence of signal intensities on the composition of the released volatile matter [18]. This observation is consistent with the results in this study, as discussed in Section 2.4. The temporal difference between the first appearance of luminescencethe and the ignition delay time determined by OH-LIF signals is quantified by $\Delta t = t_{i,LU} - t_{ign}$. Figure 13 shows PDFs of Δt for all investigated atmospheres with approximately 200 particles included in each case. The PDFs reveal similar trends for all atmospheres with peak values between 3 - 5 ms. No significant correlation between Δt and the O_2 concentration can be identified. This can be explained by the large variation of released volatile matter and soot formation of individual particles, which dominate the luminescence intensity. However, a greater Δt is observed for large particles which is probably due to the lower particle heating rate and slower devolatilization process (not shown here). The CH* and OH* signals are too weak for time-resolved particle tracking, even with an intensified camera [13]. Broad-band luminescence imaging (without intensifiers) provides higher and therefore more easily detectable signal intensities, whereas the ignition delay time is overestimated, since the luminescence intensity is only strong at high temperatures which are not present at the onset of ignition. This observation needs to be carefully considered to avoid a misinterpretation of experimental data. In this work, high sensitivity and accuracy for the ignition detection are enabled from the time-resolved OH-LIF imaging.

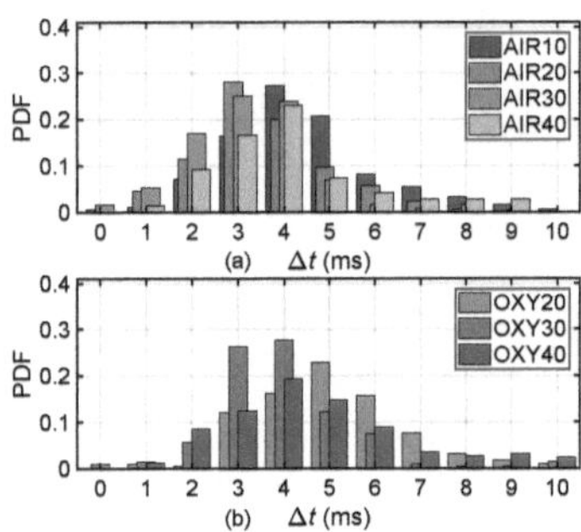

Figure 13: Comparison between LU imaging and OH-LIF with respect to the determination of the ignition delay time. $\Delta t = t_{i,LU} - t_{ign}$, where $t_{i,LU}$ corresponds to the first appearance of the detectable luminescence signal.

4.5. Ignition delay time

The ignition delay time t_{ign} is evaluated for two coal samples for different oxygen mole fractions in (a) AIR and (b) OXY conditions, as shown in Fig. 14. The time at which the particle crosses the flame front of the premixed CH_4 flame is set to $t = 0$ and characterizes the beginning of the particle heating. For each condition, more than 60 individual particles are used to determine mean values (dots) and standard deviations (bars). In general, particles with large d_p show a delayed ignition compared to smaller particles. With an increase in the O_2 mole fraction within the atmosphere, all particles show a decreasing ignition delay time. Further, the O_2 effect seems to be more significant for larger particle sizes. Comparing N_2 with CO_2 atmospheres, t_{ign} remains in a similar range when considering the same oxygen concentration. For larger particles, however, a slightly lower ignition delay time is observed in CO_2 atmospheres. The homogeneous ignition strongly depends on the gas temperature T_g and fuel mass fraction $Y_{F,0}$ of the volatile-oxidizer mixture in the vicinity of the particle. In the gas phase downstream of the particle, ignition occurs at the lean side of mixtures with increasing temperature; this has been discussed based on the temporal evolution of volumetric volatile flame structures in [20] and in other studies [44]. In the present configuration, the global gas temperatures are similar for all investigated atmospheres. However, the local gas temperature in the vicinity of particles, dominating the particle heating rate and hence the ignition, is strongly influenced by the gas composition and particle size and probably deviates from the global temperatures. A detailed discussion including a numerical analysis is provided in Section 6.

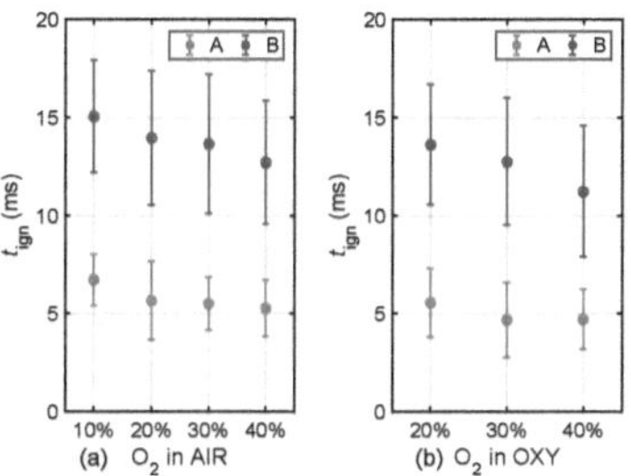

Figure 14: The ignition delay time t_{ign} in (a) AIR and (b) OXY atmospheres for differently sized particles A (mean $d_{\mathrm{p}} = 120\,\mu\mathrm{m}$) and B (mean $d_{\mathrm{p}} = 205\,\mu\mathrm{m}$).

4.6. Volatile flame duration

Figure 15 depicts the statistical results of the entire volatile flame duration t_{vol} computed from combined data from OH-LIF and LU imaging. The same data set used for the evaluation of t_{ign} is applied in this case. With an increase in particle size from 120 to 205 μm, t_{vol} rises by a factor of $2 \sim 3$. With more oxygen added into the atmosphere, t_{vol}, especially for larger particles, evidently decreases in both N_2 and CO_2 atmospheres. This indicates that the consumption rate of volatile matter is accelerated by the presence of increased amounts of oxygen. When replacing N_2 with CO_2, volatile combustion is retarded. Again, this effect is more remarkable for larger particles, while the differences for smaller particles are minor. After the onset of ignition, a gas flame rapidly spreads around the particle, while the particle continues to release volatile matter such as light hydrocarbons and tars. The volatile flame is usually considered as a typical diffusion flame in which the fuel consumption rate is controlled by molecular and thermal diffusion processes. Taking the complexity of this multi-phase phenomenon into account, effects of various parameters are hereafter discussed regarding different underlying sub-processes.

5. Numerical results

For the appropriate interpretation of experimental observations, detailed numerical simulations were performed using a combined Eulerian-Lagrangian framework,

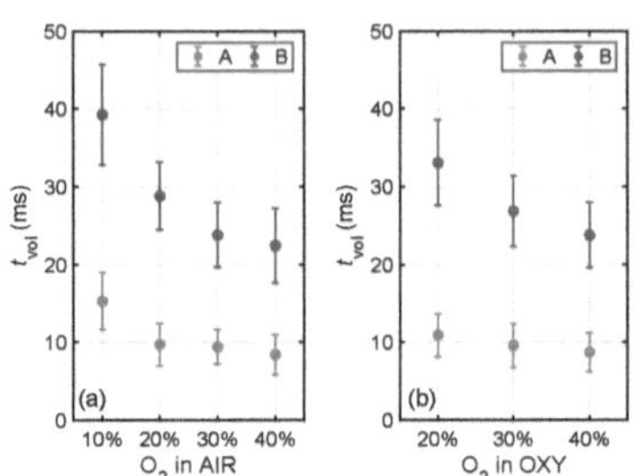

Figure 15: The volatile combustion duration t_{vol} in (a) AIR and (b) OXY atmospheres for differently sized particles A (mean $d_{\text{p}} = 120\,\mu\text{m}$) and B (mean $d_{\text{p}} = 205\,\mu\text{m}$).

as described in Section 3. The simulations provide access to essential quantities during volatile combustion not readily measurable with state-of-the-art laser diagnostics. Several cases were considered for simulation and are listed in Table 3 denoted by C_1-C_8. For a direct comparison with experiments, cases C_1-C_4 employ similar inlet conditions to the experiments and the initial U_s corresponds to the value after particles cross the flat flame at $y \approx 2\,\text{mm}$. To study the effect of the particle size, C_5 and C_6 utilize different particle diameters as supplements to the standard size of samples A and B. The slip velocities are linearly interpolated with the assumption that the particle acceleration through the premixed CH_4 flame is only a function of the particle diameter. In C_7 and C_8, U_s is increased to the value of C_1 and C_2, respectively, which enables a study of heat convection with variations in flow dynamics. For all cases, the inlet gas composition, exhaust gas temperature and velocity remain identical with the experiments, whereas U_s varies only by changing the initial particle velocity.

Table 3: Inlet parameters for all cases investigated in simulation.

Case	C_1	C_2	C_3	C_4	C_5	C_6	C_7	C_8
Atmosphere	AIR30		OXY30					
d_{p} [μm]	120	205	120	205	80	160	120	205
U_s [m/s]	0.81	1.17	0.39	0.79	0.2	0.6	0.81	1.17
T_{g} [K]	1829							

The numerical results of the ignition delay time and the volatile combustion duration are validated against the available measurements. In the simulation, t_{ign} has been quantified by taking the time at which 5% of the first peak in OH mass fraction, Y_{OH}, has been reached. This definition aims to reproduce the ignition detection of the experimental data, in which the first obvious OH signals above the background is detected. An example of the temporal change of Y_{OH} is shown in Fig. 16. This method is considered to provide accurate values to imitate the OH-LIF measurements.

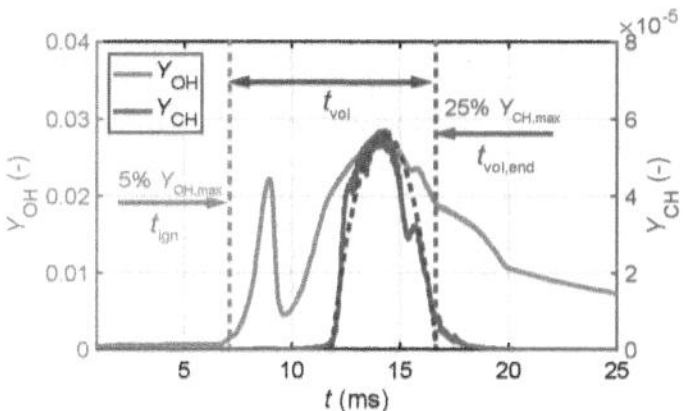

Figure 16: Definition of the ignition delay time t_{ign} and the volatile combustion duration t_{vol} in the simulations.

In Fig. 17(a), t_{ign} is compared for different particle diameters d_p. In general, the trends of delayed ignition with growing particle diameter are properly captured in accordance with experimental results. For particles with $d_p = 120\,\mu m$, the simulations are able to predict the correct t_{ign} for AIR30, while slightly over-estimating t_{ign} for OXY30. However, considering the fluctuations in the experimental data, numerical predictions remain within the error bars of experiments. Hence, it can be concluded, the simulations provide reasonable predictions in ignition delay time for smaller particles.

For larger particles, an over-prediction of the ignition delay time in the model is observed. Explanations for this deviation can be found in both measurements and simulations. In simulations, the over-prediction can be explained by Eulerian-Lagrangian modeling and the point particle approximation [44]. In the point particle approximation model, coal particles are assumed as points which represent homogeneous spheres without any temperature gradient inside the particle which release volatile matter homogeneously in all directions. Defined by the CPD model, the devolatilization occurs as the bonds in the internal structure of the coal particle break when exposed to heat at a certain particle temperature T_p which is essentially influenced by the initial particle mass and the heating rate. At a high heating rate in the order of $10^5\,K/s$ in the present work, the devolatilization has been experimentally observed as a surface phenomenon [45]. The Fourier number Fo of ~ 0.08 and ~ 0.02 are estimated for smaller and larger particles, respectively, [20] which implies the existence of evident temperature gradients in the interior of particles. This

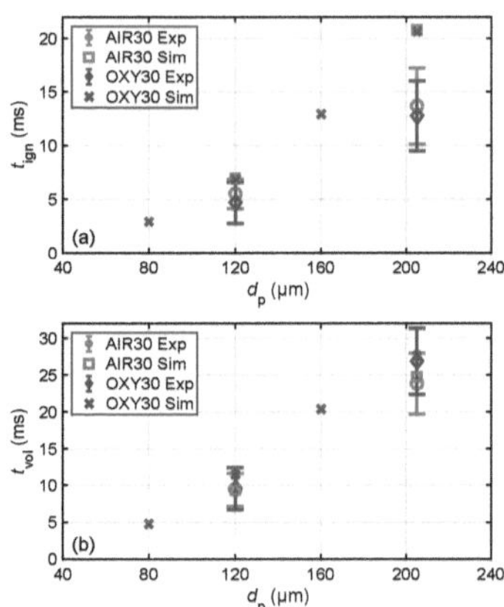

Figure 17: Comparison between numerical and experimental results with respect to t_{ign} and t_{vol} in AIR30 and OXY30 conditions, including cases from C_1 to C_6.

leads to a volatile release at elevated particle surface temperatures while the particle core is colder. When assuming a uniform particle temperature, simulations underestimate the surface temperature and thus overpredict the time of volatile release and ignition. As expected, the deviation between experiments and simulations is more distinct for larger particles, shown in Fig. 17a, probably due to the higher temperature gradient indicated by the Fourier number. The point particle approximation is considered as the major reason for the over-prediction of the ignition delay time in the present configuration with high heating rates. To improve the predictions, the temperature gradients need to be numerically considered which requires a particle-resolved simulation in future work. Additionally, in the Eulerian-Lagrangian modeling, the computational mesh resolution is limited by the particle diameter, and the steep temperature gradients in the flow cannot be accurately predicted, especially for larger particle diameters. In experiments, the determination of the ignition delay time with non-quantitative measurements is restricted by the non-linear response of OH signals to the mole fraction. Although the temperature variation is considered as negligible (see Section 4.2), interfering effects of collisional quenching and non-linearity of the image intensifier are not considered.

Additionally, the volatile combustion duration t_{vol} is validated against experimental data in Fig. 17(b). t_{vol} is defined in the simulation by the time difference between the ignition delay time and the time at which Y_{CH} decreases to 25% of its maximum value, shown in Fig. 16. This definition resembles the CH* chemiluminescence measurements in a previous study [13] which is considered to provide accurate values to imitate LU measurements. The CH mass fraction can be represented by a second-order polynomial (black dashed line in Fig. 16) centered at the maximum peak as described in [13] and [24]. From the intersection of the second-order polynomial fit with the baseline, the end of volatile combustion can be estimated. Overall good agreement for the volatile combustion duration can be found. Considering the uncertainties in both simulations and experiments, it can be concluded that the selected model can predict the physical behavior of ignition and volatile combustion of single hvb coal particles reasonably.

6. Discussion

In coal combustion, the energy conservation of coal particles can be described as a change of the particle internal energy by convective and radiative heat transfer. Hence, the particle heating rate can be considered using the following equation (adapted from [5])

$$\frac{\mathrm{d}T_\mathrm{p}}{\mathrm{d}t} = \frac{6}{d_\mathrm{p}\rho_\mathrm{p}c_\mathrm{p}}(h(T_\mathrm{g} - T_\mathrm{p}) - \sigma(\varepsilon_\mathrm{g}T_\mathrm{g}^4 - \varepsilon_\mathrm{p}T_\mathrm{p}^4)), \tag{3}$$

where T_p, ρ_p, c_p and ε_p denote the temperature, density, heat capacity and emissivity of particles in order; h is the convection coefficient of heat transfer; T_g and ε_g are the temperature and emissivity of gases, respectively, and σ is the Stefan-Boltzmann constant. The wall radiation and the heat release of solid-phase devolatilization reactions are considered negligible under the present conditions. The heating rate is strongly impacted by the particle size, thermal convection, and radiation near the solid-gas interface. In the pre-ignition stage (i.e. water evaporation and particle heating), T_g is equal to the local gas temperature close to the particle initialized by given boundary conditions, whereas increasing towards the adiabatic flame temperature T_ad (changing with atmospheres) after the onset of ignition. The heating rate is a crucial quantity to understand pulverized coal combustion for single particles. In the following, different effects on the homogeneous ignition and volatile combustion are discussed with respect to the observations noticed in experiments and simulations.

6.1. Effect of particle size

With an increasing particle diameter d_p, both t_ign and t_vol evidently increase, as shown in the experimental results of Fig. 14 and Fig. 15. In the aforementioned validation, simulations properly captured this behavior as one can observe in Fig. 17. It is then of interest to understand the effects of particle size on single particle combustion. Considering the energy balance in Eq. 3, the heating rate decreases with increasing d_p if other parameters remain unchanged. To further explain this phenomenon, the particle temperature histories of differently sized particles are numerically evaluated for cases C_3 to C_6 under the condition OXY30. Figure 18(a) shows an evidently lower slope in the rise of T_p and delayed ignition (dashed lines) for larger particle diameters. Additionally, the local gas temperature $T_\mathrm{g,local}$ is calculated by averaging T_g within a volume of $1\,\mathrm{mm}^3$ centered around the particle. In Fig. 18(b), an apparent decrease of $T_\mathrm{g,local}$ at the early heating stage is noticed. The thermal energy stored within the gas phase is responsible for the heat supply for the evaporation of moisture, increasing the particle temperature and volatile release. As heat transfer from the gas phase to the particle occurs, the gas phase temperature close to the particle is lowered, therefore convective heat transfer, which scales linearly with the temperature difference, is decreased. Due to the higher thermal inertia of larger particles, more energy is required from the gas phase for the particle heat-up, which leads to a more significant gas temperature drop as one can see in Fig. 18(b). Consequently, a lower local gas temperature leads to a further decrease of the heating rate, as indicated in Eq. 3. Moreover, the subsequent devolatilization process is essentially impacted by the initial particle heating. The volatile release rate can be depicted by the temporal evolution of the particle mass loss ratio, $m_\mathrm{p}/m_\mathrm{p,0}$, where m_p and $m_\mathrm{p,0}$ denote the instantaneous and initial particle mass respectively. Figure 18(c) indicates a temporally retarded volatile release with increasing particle size. Similar with t_ign, the increase of t_vol can be explained by (1) the longer devolatilization due to the lower heating rate as discussed above and (2) the higher volatile mass leading to an increase of the entire duration for the release of hydrocarbons and their consumption through combustion.

6.2. Effect of oxygen concentration

In experiments, it has been observed under both AIR and OXY conditions, that the ignition delay time decreases with increasing oxygen concentration as seen in Fig. 14. In other experimental studies, similar trends were observed by means of time-averaged CH* imaging measurements [13] and single-shot sooting flame imaging measurements [14]. An explanation can be found in the molecular diffusion rate, which is proportionally scaled to the concentration gradient. Thus, a higher oxygen partial pressure promotes the formation of a flammable gas mixture inducing auto-ignition. Besides, an increased O_2 content leads to a higher oxygen atom concentration close to the particle and hence further increased reactivity of fuel-oxidizer mixtures. Ignition preferentially occurs under lean conditions [20, 25, 44] for which

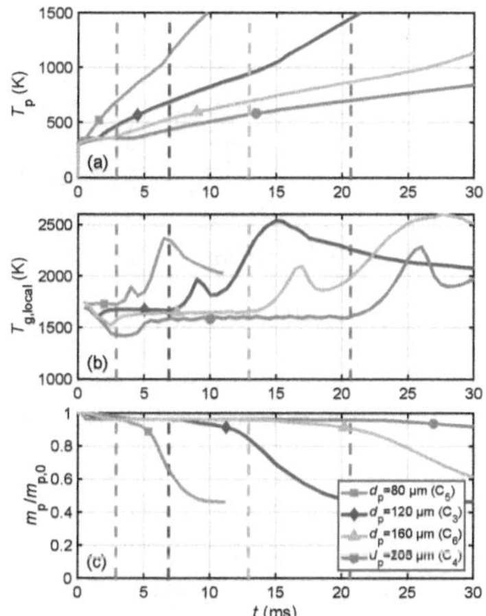

Figure 18: Effect of particle size. (a) Particle temperatures T_p, (b) local gas temperatures $T_\mathrm{g,local}$ and (c) particle mass loss $m_\mathrm{p}/m_\mathrm{p,0}$ histories for different particle sizes d_p of 80 µm (C_5), 120 µm (C_3), 160 µm (C_6), and 205 µm(C_4) in OXY30. The ignition delay time is indicated by dashed lines.

higher oxygen concentrations in the atmosphere are advantageous. An analysis of the mixture fraction field from simulations is expected to provide more insights in future investigations.

Further, increasing oxygen concentrations result in a shorter duration of volatile combustion. The same effect has been investigated using particle temperature histories measured by optical pyrometry [8, 9] or by classifying soot images to derive volatile combustion durations [14]. An accelerated devolatilization process can be argued for higher oxygen concentration with several evidences as follows. First, fuel oxidation in volatile diffusion flames benefits from faster oxygen diffusion due to a higher partial pressure. Moreover, sooting flame temperatures increase in O_2-enriched environments [8, 11, 12, 14, 15]. A higher gas temperature (1) increases the diffusivity of fuel and oxidizer and (2) increases the heating rate which expedites volatile release as well (e.g. see Eq.3). Furthermore, in the presence of more O_2, the flame size reduces [9, 12] and the reaction zone is located closer to the particle surface [18], which increases the heating rate through thermal conduction.

6.3. Effect of slip velocity

As the slip velocity, defined as the relative velocity between the particle and the gas phase, is difficult to examine experimentally, its effect is hereafter explored by means of numerical simulation. The cases C_7 and C_8 using artificially increased slip velocities are compared with the cases C_3 and C_4 with respect to t_{ign}, T_p, and $T_{g,\,local}$. Figure 19(a) shows that increased slip velocities decrease the ignition delay time. Two effects might explain this behavior:

(1) accelerated convective heating due to an increased Nusselt number Nu, and,

(2) accelerated convective heating due to a higher local gas phase temperature as the particle moves through the domain and draws energy from the gas phase.

Due to the dependency of forced convection on velocity, one can assume that an increased relative velocity between particle and gas phase results in a more efficient heat transfer to the particle. Indeed, convective heat transfer from the Eulerian to the Lagrangian phase scales linearly with the Nusselt number. In the present numerical setup, the Nusselt number is computed as

$$Nu = 2 + 0.552\sqrt{Re_p}Pr^{\frac{1}{3}}, \tag{4}$$

where $Pr = 0.7$ following calculations by Ranz and Marshall [46]. However, from Eq. 4 it can be seen, that for a realistic range of the particle Reynolds number Re_p, the effect of the slip velocity on the Nusselt number is small. Therefore, the major contribution to increased heating rates for high slip velocities leading up to combustion is the higher temperature of the gas field surrounding the particle. In the pre-ignition stage, water evaporation and particle heating causes a temperature decrease in the vicinity of the particle. With increasing slip velocities, heat transfer from the gas phase to the particle is more efficient due to the ongoing exchange of the adjacent gas phase. As heat is drawn from the surrounding gas phase, a higher slip causes a faster replacement of the cooled-down gas with hot gas upstream of the particle, therefore establishing an overall higher temperature difference which induces a faster heating of the particle. Figure 19 illustrates this effect for both slip-induced increased particle temperatures T_p in Fig. 19(b) and the simultaneous cool-down of the adjacent gas phase for high slip velocities in Fig. 19(c). Subsequently, volatile release is expedited because of higher slip velocities and therefore higher heating rates which eventually causes earlier ignition as shown in Fig. 19(a). The positive effect of slip velocity on particle ignition was observed in the previous work with a similar configuration [47]. It was concluded that the ignition delay time was longer when assuming zero slip velocity.

After reaching the maximum volatile release rate, e.g. at 15 ms for particles with a diameter of 120 μm, the majority of volatiles have already been released and the particle has lost approximately 35% of its initial mass as can be seen in Fig. 12(c). At this stage, the heat release of the surrounding volatile diffusion flame has increased the local gas temperature above its initial value. Therefore, high slip now promotes convective heat loss, which is visible in the stronger decrease of the local gas phase temperature $T_{g,local}$ after reaching its peak in Fig. 19(c). Consequently, the situation is inverted as the heating rate of particles with higher slip velocity is now lower. In addition, a higher U_s promotes gas species transport by enhancing convection, which results in faster replacement of the remaining combustible gas species fuels around the particle with the non-reactive surrounding gas. This results, amongst other effects, in an earlier decrease in the CH mass fraction and as a result in a decrease of the volatile combustion duration as one can observe in Fig. 19(a).

6.4. Effect of CO_2

In the literature, different experiments [11, 13, 14] on the single-particle scale have been conducted indicating that the replacement of N_2 with CO_2 delays particle ignition. This is usually explained based on the molar heat capacity (or so-called heat sink, ρc_p) [4, 5, 13, 14], which is sufficiently larger for CO_2 compared to N_2 (e.g. a ratio $CO_2/N_2 = 1.7$ at 1400K). In this study, we observe that the differences of the ignition delay time between N_2 and CO_2 atmospheres are minor for small particles, as can be seen in Fig. 14. For large particles with $d_p = 205$ um, t_{ign} is equivalent or even slightly decreased in the presence of CO_2. A delay in the ignition time with N_2 replaced by CO_2 is not observed in the present work. Based on this experimental observation, theoretical and numerical foundations are discussed as follows.

For auto-ignition following the classical thermal explosion problem, the ignition delay time τ_i of a reactant gas mixture can be expressed as [48]:

$$\tau_i = \frac{c_v T_0^2 / T_a}{q_c Y_{F,0} A \exp(-T_a/T_0)}, \tag{5}$$

where c_v denotes the specific heat capacity of the mixture, q_c is the reaction heat release rate, T_0 is the initial temperature, T_a is the activation temperature, $Y_{F,0}$ is the fuel mass fraction, and A is the pre-exponential factor for reaction kinetics. It is commonly argued that the higher c_v of CO_2 slows down the temperature rise and hence chemical reaction leading to a larger ignition delay time. However, it is important to note, that the impact of c_v cannot be solely considered when introducing CO_2, because the ignition delay time is a function of multiple parameters.

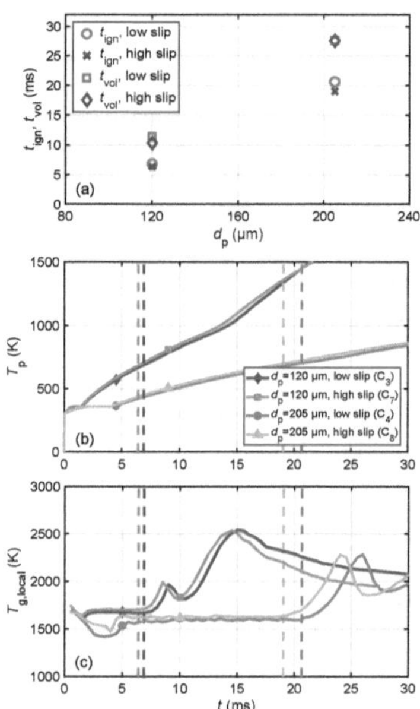

Figure 19: Effect of slip velocity. (a) Ignition delay time and volatile combustion duration, (b) particle temperature and (c) local gas temperature evolution for two different particle sizes in OXY30 condition. The cases C_3, C_4, C_7, and C_8 are included, and the corresponding t_{ign} are indicated by dashed lines in (b) and (c).

It is difficult to systematically explore the changes that CO_2 introduces by only exchanging the atmosphere, even under fully-defined conditions in the present experiments, since e.g. the flow rates for all CO_2 atmospheres are reduced to obtain gas temperature profiles similar to N_2 atmospheres. As a consequence, the particle slip velocity is lower in OXY conditions, as shown in Fig. 12 which is another parameter crucially influencing the ignition and combustion behavior of solid fuel particles. With the aid of simulations, it has been discussed above that a higher slip velocity increases heat convection and hence shortens ignition delay if other boundary conditions remain unchanged. In other words, assuming the same slip velocity as in AIR conditions, t_{ign} in OXY conditions (see Fig. 14(b)) would further decrease leading to more significant differences between N_2 and CO_2 atmospheres. In summary it can be said, that an experimental study of the influence of CO_2 always introduces changes in either thermodynamics (i.e. temperatures) or aerodynamics (i.e. slip velocities) when trying to maintain equal conditions for both AIR and OXY atmospheres. To overcome this inherent problem, this study employs a combined experimental-numerical approach.

As illustrated in Fig. 20, a reasonable comparison can only be drawn with identical slip velocities and initial gas temperatures. Here, the numerical results of cases C_1, C_2, C_7, and C_8 are compared. During the particle heating, heat transfer (convective, conductive and radiative) from the gas phase to the solid phase leads to a drop in the gas temperature in the vicinity of particles, as shown in Fig. 20(a). Considering the same amount of heat extracted, the gas temperature decreases less in CO_2 than in N_2 due to a higher c_v. For the same particle diameter, it can be seen that the local gas temperature $T_{g,local}$ is higher in the presence of CO_2 than within N_2 atmospheres, especially during the particle heating. This discrepancy is more evident for a larger particle diameter of $d_p = 205\,\mu m$ since more overall energy is required for heating. A higher local gas temperature increases the particle heating rate before ignition, as indicated in Eq. 3. In addition, the thermal conductivity of CO_2 is about 1.2 times higher than that of N_2 at 1400 K [4]. Hence, an earlier volatile release and formation of an ignitable fuel mass fraction ($Y_{F,0}$) can be expected. This is verified by computing the maximum fuel mass fractions $Y_{C_2H_2}+Y_{CH_4}$ in the vicinity of the particle as a representative $Y_{F,0}$ for volatile combustion. In Fig. 20(b), an early increase of fuel mass fraction is noticeable in the CO_2 atmosphere, which also implies an early formed flammable gas mixture. Equation 5 indicates that the gas temperature T_0 (here equivalent to $T_{g,local}$) and $Y_{F,0}$ have opposite effects compared with c_v or ρc_v on the ignition delay time. For example, Khatami et al. [15] explained in DTF measurements that the retarded ignition in CO_2 is mainly attributed to the lower ambient temperature compared to that in N_2 in their facility. Since differences of $T_{g,local}$ at ignition are minor in this study, as indicated by Fig. 20(a), the fuel mass fraction is considered as a dominant effects probably inducing early ignition. In summary, based on the present study which has been performed at an initial gas temperature of around 1800 K, combining experiments and simulations, the replacement of N_2 with CO_2 does not necessarily result in a delayed ignition with the same initial global gas temperature. In general, CO_2 introduces competing effects on the ignition delay time via the changes of the local gas temperature and the heat capacity. It is important to emphasize that auto-ignition is a local phenomenon, which requires an examination of the possible influence factors (i.e. in Eq. 5) in a relevant domain. Although it is strongly impacted by global boundary conditions, the local heating conditions and gas mixture fraction are dominating factors and need to be considered carefully.

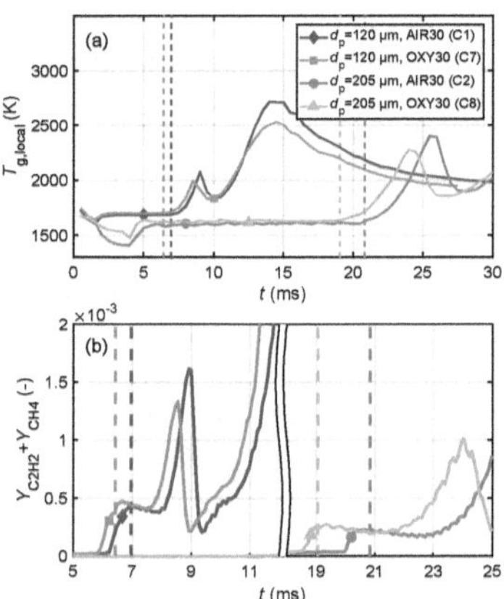

Figure 20: Effect of CO_2. (a) Local gas temperature $T_{g,\,local}$ histories and (b) the temporal change of the maximum fuel mass fraction (represented by $Y_{C_2H_2} + Y_{CH_4}$) of two different particle sizes during the ignition under AIR30 and OXY30 conditions. The cases C_1, C_2, C_7, and C_8 are compared, and the corresponding t_{ign} are indicated by dashed lines.

Previous experimental [14, 15] and numerical [22, 49] investigations have reported convincing evidences of a longer volatile combustion duration in CO_2 atmospheres. In accordance with literature, in Fig. 15 and Fig. 17, the increase of the volatile combustion duration is observed in both experimental and numerical data, which appears more clearly for larger particles. Considering the volatile flame as a typical diffusion flame, there are several explanations. Above all, gas diffusion is retarded by replacing N_2 with CO_2. The diffusivity of O_2 into CO_2 is 0.75 times that into N_2 at 1400 K and light hydrocarbons like CH_4 also show a lower diffusivity in CO_2. In addition, the peak flame temperatures are lower in CO_2 due to the high heat capacity. This effect has been reported in [14, 15] and is also evident according to our results in Fig. 20. Moreover, the O_2 diffusivity is a function of temperature which decreases further in CO_2 atmospheres. Although the effect might be minor, the entire volatile yield is reported to be higher in CO_2 atmospheres. In conclusion, a retarded gas diffusion and an overall slowed fuel consumption rate can be expected when replacing N_2 with CO_2.

7. Conclusions

The present work reports a detailed investigation of single particle combustion of a Colombian hvb coal under conventional and oxy-fuel atmospheres using a combined experimental-numerical approach. The laminar flow reactor was comprehensively characterized by a series of experiments e.g. with PIV and quantitative OH-LIF, providing fully defined boundary conditions of e.g. inlet gas compositions, outlet gas temperature and velocity profiles, particle heating points (i.e. the position of the flame front), and particle size and shape. Two particle samples A (mean $d_p = 120\,\mu m$) and B (mean $d_p = 205\,\mu m$) were investigated in four N_2 atmospheres (containing 10vol% - 40vol% O_2) and three CO_2 (containing 20vol% - 40vol% O_2) atmospheres. Multi-parameter optical measurements enabled the simultaneous detection of the ignition delay time t_{ign} and the volatile combustion duration t_{vol} for each individual coal particle. Time-resolved OH-LIF revealed a high sensitivity and accuracy for ignition detection, while broad-band luminescence imaging over-predicted homogeneous ignition with a delay of $3 \sim 5\,ms$ on average. Numerical simulations performed in an Eulerian-Lagrangian framework were validated against experimental results and further provided insights into multi-phase heat transfer and volatile release by computing essential quantities of e.g. T_p, $T_{g,local}$, $m_p/m_{p,0}$ and Y_F. By analyzing experimental and numerical observations and including theoretical evidences, the effects of particle size d_p, O_2 enrichment, slip velocity U_s, and CO_2 addition were discussed with regard to t_{ign} and t_{vol}. Several conclusions are summarized as follows:

(1) Both t_{ign} and t_{vol} significantly increase with the particle diameter d_p. The retarded ignition is attributed to a decreased particle heating rate, which is associated with a larger d_p and lower $T_{g,local}$ induced by heat transfer between gas phase and particle. The increased volatile combustion duration mainly results from a higher volatile content stored in larger particles which eventually extends the devolatilization period, whereas the different particle heating rates before the onset of ignition might have minor influences on t_{vol}.

(2) Both t_{ign} and t_{vol} decrease with an increasing O_2 concentration in all investigated atmospheres, with a more pronounced effect for larger particles. The higher partial pressure of O_2 accelerates the formation of a flammable gas mixture (i.e. shortens t_{ign}) and raises the fuel consumption rate (i.e. shortens t_{vol}) by faster molecular diffusion. For t_{vol}, the O_2-dependent flame temperature and flame size are considered to be of secondary importance.

(3) Higher slip velocity accelerates the heat convection between the gas phase and the particle due to a faster replacement of the cooled down gas around the par-

ticle with hot gas upstream of the particle, which eventually reduces t_{ign}. After releasing the majority of volatiles, heat loss and molecular transport by convection close to the particle are enhanced by an increased U_s, and accordingly, t_{vol} shows a slight tendency to decrease due to a higher fuel consumption rate.

(4) With an identical U_s and global T_g, CO_2 mitigates the decrease in $T_{g,local}$ caused by evaporation and particle heating in the pre-ignition stage due to its higher heat capacity compared with N_2. This subsequently results in an early release of light hydrocarbons and flammable gas mixtures promoting auto-ignition. This effect is more significant for larger particles, which is seen in both simulations and experiments. On the other hand, volatile combustion is retarded by replacing N_2 with CO_2, which is explained by the lower diffusivity of both O_2 and light hydrocarbons within CO_2.

Acknowledgements

The authors kindly acknowledge financial support through Deutsche Forschungsgemeinschaft (DFG) – Projektnummer 215035359 – TRR 129 for its support through CRC/Transregio 129 "Oxy-flame: development of methods and models to describe solid fuel reactions within an oxy-fuel atmosphere." Gefördert durch die Deutsche Forschungsgemeinschaft (DFG) - Projektnummer 215035359 - TRR 129.

References

[1] BP, Statistical review of world energy 2020 : A pivotal moment (2020).
URL www.bp.com/en/global/corporate/energy-economics/statistical-review-of

[2] T. F. Wall, Combustion processes for carbon capture, Proceedings of the Combustion Institute 31 (1) (2007) 31–47. doi:10.1016/j.proci.2006.08.123.

[3] T. Wall, Y. Liu, C. Spero, L. Elliott, S. Khare, R. Rathnam, F. Zeenathal, B. Moghtaderi, B. Buhre, C. Sheng, R. Gupta, T. Yamada, K. Makino, J. Yu, An overview on oxyfuel coal combustion—state of the art research and technology development, Chemical Engineering Research and Design 87 (8) (2009) 1003–1016. doi:10.1016/j.cherd.2009.02.005.

[4] M. B. Toftegaard, J. Brix, P. A. Jensen, P. Glarborg, A. D. Jensen, Oxy-fuel combustion of solid fuels, Progress in Energy and Combustion Science 36 (5) (2010) 581–625. doi:10.1016/j.pecs.2010.02.001.

[5] L. Chen, S. Z. Yong, A. F. Ghoniem, Oxy-fuel combustion of pulverized coal: Characterization, fundamentals, stabilization and cfd modeling, Progress in Energy and Combustion Science 38 (2) (2012) 156–214. doi:10.1016/j.pecs.2011.09.003.

[6] S. Li, Y. Xu, Q. Gao, Measurements and modelling of oxy-fuel coal combustion, Proceedings of the Combustion Institute 37 (3) (2019) 2643–2661. doi:10.1016/j.proci.2018.08.054.

[7] A. Williams, M. Pourkashanian, J. M. Jones, Combustion of pulverised coal and biomass, Progress in Energy and Combustion Science 27 (6) (2001) 587–610. doi:10.1016/S0360-1285(01)00004-1.

[8] L. D. Timothy, A. F. Sarofim, J. M. Béer, Characteristics of single particle coal combustion, Symposium (International) on Combustion 19 (1) (1982) 1123–1130. doi:10.1016/S0082-0784(82)80288-9.

[9] L. D. Timothy, D. Froelich, A. F. Sarofim, J. M. Béer, Soot formation and burnout during the combustion of dispersed pulverized coal particles, Symposium (International) on Combustion 21 (1) (1988) 1141–1148. doi:10.1016/S0082-0784(88)80345-X.

[10] P. A. Bejarano, Y. A. Levendis, Single-coal-particle combustion in o2/n2 and o2/co2 environments, Combustion and Flame 153 (1-2) (2008) 270–287. doi:10.1016/j.combustflame.2007.10.022.

[11] R. Khatami, C. Stivers, Y. A. Levendis, Ignition characteristics of single coal particles from three different ranks in o2/n2 and o2/co2 atmospheres, Combustion and Flame 159 (12) (2012) 3554–3568. doi:10.1016/j.combustflame.2012.06.019.

[12] R. Khatami, Y. A. Levendis, M. A. Delichatsios, Soot loading, temperature and size of single coal particle envelope flames in conventional- and oxy-combustion conditions (o 2 /n 2 and o 2 /co 2), Combustion and Flame 162 (6) (2015) 2508–2517. doi:10.1016/j.combustflame.2015.02.020.

[13] A. Molina, C. R. Shaddix, Ignition and devolatilization of pulverized bituminous coal particles during oxygen/carbon dioxide coal combustion, Proceedings of the Combustion Institute 31 (2) (2007) 1905–1912. doi:10.1016/j.proci.2006.08.102.

[14] C. R. Shaddix, A. Molina, Particle imaging of ignition and devolatilization of pulverized coal during oxy-fuel combustion, Proceedings of the Combustion Institute 32 (2) (2009) 2091–2098. doi:10.1016/j.proci.2008.06.157.

[15] R. Khatami, C. Stivers, K. Joshi, Y. A. Levendis, A. F. Sarofim, Combustion behavior of single particles from three different coal ranks and from sugar cane bagasse in o2/n2 and o2/co2 atmospheres, Combustion and Flame 159 (3) (2012) 1253–1271. doi:10.1016/j.combustflame.2011.09.009.

[16] J. Köser, L. G. Becker, N. Vorobiev, M. Schiemann, V. Scherer, B. Böhm, A. Dreizler, Characterization of single coal particle combustion within oxygen-enriched environments using high-speed oh-plif, Applied Physics B 121 (4) (2015) 459–464. doi:10.1007/s00340-015-6253-3.

[17] J. Köser, L. G. Becker, A.-K. Goßmann, B. Böhm, A. Dreizler, Investigation of ignition and volatile combustion of single coal particles within oxygen-enriched atmospheres using high-speed oh-plif, Proceedings of the Combustion Institute 36 (2) (2017) 2103–2111. doi:10.1016/j.proci.2016.07.083.

[18] J. Köser, T. Li, N. Vorobiev, A. Dreizler, M. Schiemann, B. Böhm, Multi-parameter diagnostics for high-resolution in-situ measurements of single coal particle combustion, Proceedings of the Combustion Institute 37 (3) (2019) 2893–2900. doi:10.1016/j.proci.2018.05.116.

[19] T. Li, C. Geschwindner, J. Köser, M. Schiemann, A. Dreizler, B. Böhm, Investigation of the transition from single to group coal particle combustion using high-speed scanning oh-lif and diuse backlight-illumination, Proceedings of the Combustion Institute (2020). doi:10.1016/j.proci.2020.06.314.

[20] T. Li, M. Schiemann, J. Köser, A. Dreizler, B. Böhm, Experimental investigations of single particle and particle group combustion in a laminar flow reactor using simultaneous volumetric oh-lif imaging and diffuse backlight-illumination, Renewable and Sustainable Energy Reviews (2020).

[21] D. M. Grant, R. J. Pugmire, T. H. Fletcher, A. R. Kerstein, Chemical model of coal devolatilization using percolation lattice statistics, Energy and Fuels 3 (2) (1989) 175–186. doi:10.1021/ef00014a011.

[22] S. Farazi, A. Attili, S. Kang, H. Pitsch, Numerical study of coal particle ignition in air and oxy-atmosphere, Proceedings of the Combustion Institute 37 (3) (2019) 2867–2874.

[23] S. Farazi, J. Hinrichs, M. Davidovic, T. Falkenstein, M. Bode, S. Kang, A. Attili, H. Pitsch, Numerical investigation of coal particle stream ignition in oxy-atmosphere, Fuel 241 (2019) 477–487.

[24] Y. Liu, M. Geier, A. Molina, C. R. Shaddix, Pulverized coal stream ignition delay under conventional and oxy-fuel combustion conditions, International Journal of Greenhouse Gas Control 5 (2011) S36–S46.

[25] G. Tufano, O. Stein, B. Wang, A. Kronenburg, M. Rieth, A. Kempf, Coal particle volatile combustion and flame interaction. part i: Characterization of transient and group effects, Fuel 229 (2018) 262–269.

[26] G. Tufano, O. Stein, B. Wang, A. Kronenburg, M. Rieth, A. Kempf, Coal particle volatile combustion and flame interaction. part ii: Effects of particle reynolds number and turbulence, Fuel 234 (2018) 723–731.

[27] T. Maffei, R. Khatami, S. Pierucci, T. Faravelli, E. Ranzi, Y. A. Levendis, Experimental and modeling study of single coal particle combustion in O_2/N_2 and Oxy-fuel (O_2/CO_2) atmospheres, Combustion and Flame 160 (11) (2013) 2559–2572.

[28] G. L. Tufano, O. T. Stein, A. Kronenburg, A. Frassoldati, T. Faravelli, L. Deng, A. M. Kempf, M. Vascellari, C. Hasse, Resolved flow simulation of pulverized coal particle devolatilization and ignition in air- and O_2/CO_2-atmospheres, Fuel 186 (2016) 285–292.

[29] S. Jimenez, C. Gonzalo-Tirado, Properties and relevance of the volatile flame of an isolated coal particle in conventional and oxy-fuel combustion conditions, Combustion and Flame 176 (2017) 94–103.

[30] L. Cai, C. Zou, Y. Guan, H. Jia, L. Zhang, C. Zheng, Effect of steam on ignition of pulverized coal particles in oxy-fuel combustion in a drop tube furnace, Fuel 182 (2016) 958–966. doi:10.1016/j.fuel.2016.05.083.

[31] J. Heinze, U. Meier, T. Behrendt, C. Willert, K.-P. Geigle, O. Lammel, R. Lückerath, Plif thermometry based on measurements of absolute concentrations of the oh radical, Zeitschrift für Physikalische Chemie 225 (11-12) (2011) 1315–1341. doi:10.1524/zpch.2011.0168.

[32] D. G. Goodwin, H. K. Moffat, R. L. Speth, Cantera: An object-oriented software toolkit for chemical kinetics, thermodynamics, and transport processes, http://www.cantera.org, version 2.1.0 (2015). doi:10.5281/zenodo.170284.

[33] J. Luque, D. R. Crosley, Lifbase version 2.1. 1, SRI International, Menlo Park, CA (1999).

[34] R. Sadanandan, W. Meier, J. Heinze, Experimental study of signal trapping of oh laser induced fluorescence and chemiluminescence in flames, Applied Physics B 106 (3) (2012) 717–724.

[35] J. Pareja, A. Johchi, T. Li, A. Dreizler, B. Böhm, A study of the spatial and temporal evolution of auto-ignition kernels using time-resolved tomographic oh-lif, Proceedings of the Combustion Institute 37 (2) (2019) 1321–1328. doi:10.1016/j.proci.2018.06.028.

[36] T. Li, B. Zhou, J. H. Frank, A. Dreizler, B. Böhm, High-speed volumetric imaging of formaldehyde in a lifted turbulent jet flame using an acousto-optic deflector, Experiments in Fluids 61 (4) (2020) 2903. doi:10.1007/s00348-020-2915-y.

[37] W. A. Sirignano, Fluid Dynamics and Transport of Droplets and Sprays, 2nd Edition, Vol. 1, 2010. doi:10.1017/CBO9781107415324.004.

[38] A. Attili, F. Bisetti, M. E. Mueller, H. Pitsch, Formation, growth, and transport of soot in a three-dimensional turbulent non-premixed jet flame, Combustion and Flame 161 (7) (2014) 1849 – 1865. doi:https://doi.org/10.1016/j.combustflame.2014.01.008.

[39] Y. Bai, K. Luo, K. Qiu, J. Fan, Numerical investigation of two-phase flame structures in a simplified coal jet flame, Fuel 182 (2016) 944–957. doi:10.1016/j.fuel.2016.05.086.

[40] L. Cai, S. Kruse, D. Felsmann, H. Pitsch, A methane mechanism for oxy-fuel combustion: Extinction experiments, model validation, and kinetic analysis, FLOW TURBULENCE AND COMBUSTION (2020).

[41] O. Desjardins, G. Blanquart, G. Balarac, H. Pitsch, High order conservative finite difference scheme for variable density low mach number turbulent flows, Journal of Computational Physics 227 (15) (2008) 7125–7159.

[42] S. Farazi, M. Sadr, S. Kang, M. Schiemann, N. Vorobiev, V. Scherer, H. Pitsch, Resolved simulations of single char particle combustion in a laminar flow field, Fuel 201 (2017) 15 – 28. doi:https://doi.org/10.1016/j.fuel.2016.11.011.

[43] B. Goshayeshi, J. Sutherland, A comparison of various models in predicting ignition delay in single-particle coal combustion, Combustion and Flame 161 (7) (2014). doi:10.1016/j.combustflame.2014.01.010.

[44] M. Vascellari, H. Xu, C. Hasse, Flamelet modeling of coal particle ignition, Proceedings of the Combustion Institute 34 (2) (2013) 2445-2452.

[45] R. Rajasegar, D. C. Kyritsis, Experimental investigation of coal combustion in coal-laden methane jets, Journal of Energy Engineering 141 (2) (2015). doi:10.1061/(ASCE)EY.1943-7897.0000228.

[46] W. Ranz, W. R. Marshall, et al., Evaporation from drops, Chem. eng. prog 48 (3) (1952) 141-146.

[47] A. Attili, P. Farmand, C. Schumann, S. Farazi, B. Böhm, T. Li, C. Geschwindner, J. Köser, A. Dreizler, H. Pitsch, Numerical simulations and experiments of ignition of solid particles in a laminar burner - effects of slip velocity and particle swelling, Flow Turbulence and Combustion Under Review (2020).

[48] C. K. Law, Combustion Physics, Cambridge University Press, Cambridge, 2006. doi:10.1017/CBO9780511754517.

[49] A. A. Bhuiyan, J. Naser, Thermal characterization of coal/straw combustion under air/oxy-fuel conditions in a swirl-stabilized furnace: A cfd modelling, Applied Thermal Engineering 93 (2016) 639-651.

Appendix E (Paper V):
Volumetric measurements of single particle combustion

Experimental investigations of single particle and particle group combustion in a laminar flow reactor using simultaneous volumetric OH-LIF imaging and diffuse backlight-illumination

Li, Tao[a],*, Schiemann, Martin[b], Köser, Jan[a], Dreizler, Andreas[a] and Böhm, Benjamin[a]

[a]*Institute Reactive Flows and Diagnostics, Technical University of Darmstadt, 64287 Darmstadt, Germany*
[b]*Department of Energy Plant Technology, Ruhr-University Bochum, 44801 Bochum, Germany*

ARTICLE INFO

Keywords:
Single particles
Particle groups
High-volatile bituminous coal
Ignition and volatile flame
Volumetric OH-LIF and DBI

ABSTRACT

The volatile combustion of Colombian high-volatile bituminous coal was experimentally studied in a laboratory laminar flow reactor. The volatile flames corresponding to the single particle and particle group combustion were visualized using non-intrusive multi-parameter optical diagnostics. In the present study, high-speed laser-induced fluorescence of OH radicals (OH-LIF) was applied to study igniting particles by temporally tracking OH-LIF signals in the gas-phase flame. A novel acousto-optic deflector combined with a 10 kHz dye laser was employed for laser scanning through a probe volume with a thickness of a few millimeters. The three-dimensional OH-LIF signals were used to reconstruct the volatile flame structures of burning particles. Simultaneously, diffuse backlight-illumination (DBI) is implemented to measure the size and the spatial distribution of particles to distinguish between single and group particle combustion. For single particles, starting from the onset of ignition, the OH-LIF intensity reaches its maximum within several milliseconds, which is temporally resolved by employing a laser scanning system. The gas-phase ignition starts downstream of the particles. As the particle size increases, the flame stand-off distance increases, whereas the ratio of the flame stand-off distance and the particle diameter decreases, which ranges from 2 to 4 for the coal particles investigated. For particle groups, the flame topology is evaluated for individual reconstructions with different particle number densities (PND). As the PND increases, the volatile flames are pushed outwards to the boundary of particle clouds and a non-flammable region emerges in the center of volatile flames. Soot formation is observed and becomes increasingly intensive as the PND increases.

Highlights

- High-resolution optical techniques provide novel data for pulverized fuel modeling

- Volumetric LIF temporally and spatially resolves the ignition and volatile combustion processes

- Gas-phase ignition of single particles starts downstream of the particles

- Particle-flame interaction impacts the stand-off distance and intensities for single particle volatile flames

- 3D flame topology changes significantly from single to group particle combustion due to particle-to-particle interaction

Word Count

Total word count 9516 = 247 (Abstract) +2341 (Introduction) + 2045 (Experimental Methods) +4124 (Results and Discussions) + 759 (Conclusion)

*Corresponding author
✉ tao.li@rsm.tu-darmstadt.de (L. Tao)
ORCID(s): 0000-0001-8942-6849 (L. Tao)

Nomenclature

Greek letters			Abbreviations	
Λ	acoustic wavelength (nm)		AOD	acoustic-optic deflector
λ	optical wavelength (nm)		CFD	computational fluid dynamics
v	acoustic velocity (m/s)		DBI	diffuse backlight-illumination
θ_B	Bragg angle (mrad)		DNS	direct numerical simulation
θ_D	deflection angle (mrad)		DOF	depth of field
θ_i	incident angle (mrad)		DSLR	digital single-lens reflex
θ_{max}	maximum deflection angle (mrad)		FOV	field of view
θ_{min}	minimum deflection angle (mrad)		FWHM	full width at half maximum
θ_{scan}	scan angle (mrad)		GPC	group particle combustion
α	thermal diffusivity		hvb	high volatile bituminous
τ	residence time		ICCD	intensified charge coupled device
			LES	large eddy simulation
Symbols			LIF	laser-induced fluorescence
d_P	equivalent particle diameter (µm)		LRF	laminar flow reactor
d	spot diameter (µm)		pf	pulverized fuel
f	optical frequency (MHz)		PIV	particle imaging velocimetry
f_L	focal length (mm)		PND	particle number density
f_c	acoustic center frequency (MHz)		PDF	probability density function
f_{min}	minimum acoustic frequency (MHz)		RANS	Reynolds-Averaged-Navier-Stokes
f_{max}	maximum acoustic frequency (MHz)		RF	radio frequency
Fo	Fourier number		SPC	single particle combustion
I_{flame}	OH-LIF intensity in flame			
$I_{max,flame}$	maximum OH-LIF intensity in flame			
G	group number			
L	interaction length (nm)			
L_x	inter-particle distance (µm)			
r_P	equivalent particle radius (µm)			
r_{flame}	flame stand-off distance (µm)			
Re_P	particle Reynolds number			
R_c	column or cylinder radius			
T_{gas}	gas temperature			

1. Introduction

Even in recent studies on climate change mitigation, coal combustion in combination with carbon capture is considered as part of the furture global energy supply [1]. A possible furture power plant technology for the use of coal as a fuel is oxy-fuel combustion [2], which has been demonstrated on a technical scale but can still benefit from deeper knowledge of combustion details. Techniques for experimental ignition tests are manifold, and comprehensive overviews on the basics were provided early on by Essenhigh et al. [3] and Annamalai et al. [4, 5]. Drop tube and enclosed laminar flow reactors (often: entrained flow reactors) are only two of many possible experimental enviroments, but for brevity only literature which is closely related to these types of setups is mentioned.

As the ignition of pulverized fuel (pf) particles and particle groups is characterized by small-scale phenomena, non-invasive optical methods allow to study these without influencing the physico-chemical processes. Table 1 lists recent work of optical measurements of the ignition of particles groups within laminar flow reactors. Although the change to renewable fuels is necessary, only a few of these studies investigate biomass ignition [6]. A second group of studies focusses on oxy-fuel combustion [7, 8, 9, 10]. All other studies focus on N_2-containing atmospheres, which are denoted as air in Table 1 for simplicity. The influence of water vapor concentration was studied in [11], which reflects the differences between dry and wet flue gas recirculation in oxy-fuel boilers [12] and also provides insights into the altered ignition behaviour when water is used for steam-moderated oxy-fuel combustion [13]. One of the listed studies presents data from relatively oxygen-lean atmospheres with different oxygen concentrations from 5% to 20% [9], which is motivated by the idea of MILD combustion, but it must be noted that the particle concentration or number density

group ignition. In contrast to all other studies, the Hencken burner in [14] allows the exposure of particles to a high-temperature reducing atmosphere before they are exposed to oxygen-rich hot gas. This represents the situation on large scale swirl burners, where internal flue gas recirculation might cause this scenario.

Different approaches are used to characterize the ignition of particle groups. The studies in Table 1 all use imaging techniques which are below the spatial resolution that would be required to resolve single particles. The measurement is therefore based on temporal integration of the particle signal into a streak (see the comparison of particle clouds in Fig. 1(b)). It is obvious that the axial intensity increases in the flow direction, as the particles ignite successively and contribute to thermal emissions. Depending on the applied camera or detector, different integration schemes were applied. Adeosun et al. used the average of 6000 high-speed video images to determine the ignition point from the increasing thermal emissions caused by combustion of volatiles and char [14]. Sarroza et al. used averaged images and determined the ignition point of the particle streak to be the position where the intensity had reached 12% of the maximum intensity [6]. A 10% threshold of the maximum intensity was similarly employed by several authors. Liu et al. used a summation of images from an ICCD camera, which was equipped with a narrow-band CH* filter (commenly used for flame front detection in gas flames), and kept the camera shutter open to integrate the signal for at least a few seconds. A 50% intensity criterion was used to characterize ignition delay times in this study [10]. A threshold of 10% of the peak intensity was chosen by Yuan et al. [7, 15, 16] and Zeng et al. [9]. Yuan et al. used both thermal emission [15] and CH* in their studies [7, 16], in the former case the 10% criterion was supplemented by the time which the intensity took to reach the 20% and 50% levels, which according to the authors gives better insight into the ignition progress. In other works, the ignition time scale was defined as the time when a majority or significant fraction of the particles was ignited [8, 9].

The particle number density PND or particle concentration is the major parameter to characterize particle group ignition regimes. The literature in Table 1 shows some variations regarding this parameter; in general, the range of particle number density or volume fraction is suitable for pulverized fuel ignition studies. Annamalai et al. reported a typical PND range of 0.5 - $5 \times 10^9 \, \mathrm{m}^{-3}$ in the feed pipe of pulverized coal burners [5]. In the given studies, only Liu et al. [10] and Prationo et al. [8] reported the concentration as the number of particles/volume in the carrier gas (3×10^8 - $6 \times 10^{10} \, \mathrm{m}^{-3}$ and 1×10^7 - $1 \times 10^9 \, \mathrm{m}^{-3}$), while all other studies only provide a mass flow rate of particles and calculate the volume fraction for the cold carrier flow based on the flow rate of the carrier gas, but do not explicitly address the fuel density used. For example, in the study which used biomass [6], a severe difference in the particles' gravimetric density between coal and biomass would directly affect the volume fraction and particle-to-particle distance.

The experimental approach using lower spatial resolution limits the comparability between the present and literature. The impact of particle concentration on ignition is clearly visible in most of the mentioned experiments. Liu et al. [10] , who studied a wide range of particle concentrations, reported a tendency for slightly accelerated ignition in the linear particle streak for concentrations rising to 2 - $3 \times 10^9 \, \mathrm{m}^{-3}$, followed by increased ignition delay times for higher concentrations. Within the studied size range (min: 54 - $74 \, \mu\mathrm{m}$, max: 106 - $125 \, \mu\mathrm{m}$), larger particles caused a slight shift of this minimum igniton delay time to smaller number densities.

Experimental studies can not only be used for phenomenological investigations, but also to derive models and to develop or validate more complex numerical simulation tools, one of these being Computational Fluid Dynamics (CFD). While Reynolds-Averaged-Navier-Stokes (RANS) simulations have previously been state-of-the-art [17, 18], Large-Eddy Simulations (LES) and Direct Numerical Simulations (DNS) have become tools which can provide a deeper understanding of combustion processes on the particle level. Knappstein et al. [19] investigated the application of tabulated chemistry for LES simulations and compared their single particle simulation to experimental OH profiles obtained from a setup similar to the one described in Section 2. The next step in simulation complexitiy, namely moving towards DNS-like simulations, often restricts the computational domain to dimensions of only a few millimetres. Further, relevant to the present experimental investigation are the numerical studies which examine L_x/d_P, the ratio of the inter-particle distance and the particle diameter, and the particle Reynolds number Re_P. Sayadi et al. [20] chose a fixed range of $L_x/d_\mathrm{P} \leqslant 9$ and $Re_\mathrm{P} \leqslant 8$, but their study was restricted to only har combustion. In a subsequent work by Farazi et al. [21], ignition of coal particles using detailed gas-phase chemistry and devolatilization rates from the chemical percolation devolatilization (CPD) model [22] was applied to simulations with boundary conditions adapted to match the work of Liu et al. [10]. The propagation of flames through the particle groups was studied based on the heat release, temperature profile, O_2 and C_2H_2 concentrations. The results clearly replacedshowedshow the influence of PND on the ignition mode. as a PND above $1.25 \times 10^8 \, \mathrm{m}^{-3}$ caused increasing ignition delay times for different sets

Table 1
Overview on particle streak ignition studies.

Refs	Optical Devices			Flow Conditions		Fuel	Atmosphere/ Variations	
	DSLR	High-speed/ ICCD	OH-PLIF	laminar	turbulent			
[6]	no	yes	no	Drop tube furnace	-	anth/bit/lig/ 4 biomass	Air	Co-firing/ coal-biomass
[7]	yes	yes	no	Hencken	-	lig	Air/ T_{gas}, O_2	CH* intensity
[8]	yes	yes	no	McKenna	-	lig/sub-bit/bit	Air, oxy-fuel/ O_2, H_2O_{vap}	-
[9]	yes	no	no	Hencken	-	sub-bit	Air, oxy-fuel/ O_2	MILD
[10]	no	yes	no	Hencken	-	3 hvb/sub-bit	Air, oxy-fuel/ O_2	CH* intensity
[14]	no	yes	no	Hencken	-	sub-bit	Air/O_2	Reducing/oxidizing transition
[15]	yes	yes	no	Hencken		lig/char	Air, oxy-fuel, T_{gas}, O_2	-
[16]	no	yes	no	Hencken	-	lig/2 bit	Air/ T_{gas}, O_2	-
[23]	yes	yes	no	-	Hencken	bit/lig	Air/O_2	-
[24]	no	no	yes	Hencken	-	2 bit	Air/ Re_{jet}	-

Notes: bit = bituminous coal; hvb = high-volatile bituminous coal; lig = lignite coal; anth = anthracite; DSLR = digital single-lens reflex; ICCD = intensified charge coupled device.

Regarding laser diagnostics, OH-LIF was successfully applied to study the single particle ignition and temporal evolution of volatile flames in a previous work by the authors of the present study [25]. The stand-off distance of volatile flames has also beeninvestigated using high-speed OH-LIF [26]. Further studies combined OH-LIF with luminescence imaging and high-resolution diffusive backlight-illumination (DBI) to obtain correlated statistics between ignition, devolatilization duration, and particle size [27]. The inherent limitation of two-dimensional approaches is the deficient out-of-plane information which frequently leads to incorrect interpretation of measurement data [28]. For a better understanding of multi-dimensional transient phenomena in group particle combustion, the extension of planar single-shot measurements to spatially and temporally resolved measurements is desired. Recently, approaches using multiple laser sheets have been proposed to be aneffective methods for performing volumetric laser experiments. By sweeping a single laser sheet rapidly through a measurement volume, multiple images can be recorded at different out-of-plane locations, which enable volumetric reconstruction. Previous optical measurements with kHz sampling rates addressed spray visualization [29], mixture fraction visualization [30], flame front, and reaction zone detection [31, 32], employing mechanical rotating mirrors, i.e. oscillating mirrors and polygonal mirrors. An intrinsic limitation of mirror-based scanners is the high inertia and instability of the moving mechanical parts, which restrict the accuracy and precision of laser beam positions, especially at high rotational speeds. In addition, laser safety issues need to be considered [32] when using such mechanical scanners. An alternative to the mirror-based scanner is the optical solid state deflector, which relies on the acousto-optic effect. An acousto-optic deflector (AOD) does not contain any moving parts and thus presents advantages over mechanical scanners in terms of scan frequency, accuracy, precision, and spatial resolution [33]. The capabilities of AOD for laser diagnostics in combustion applications have been demonstrated on flame structure visualization of a turbulent lifted jet flame by means of Mie scattering [34] and LIF imaging [35], and by in-cylinder flow measurements in an internal combustion (IC) engine via multi-plane particle image velocimetry (PIV) [36]. However, to the best of the authors' knowledge, no laser scanning measurements have been performed to investigate coal volatile combustion.

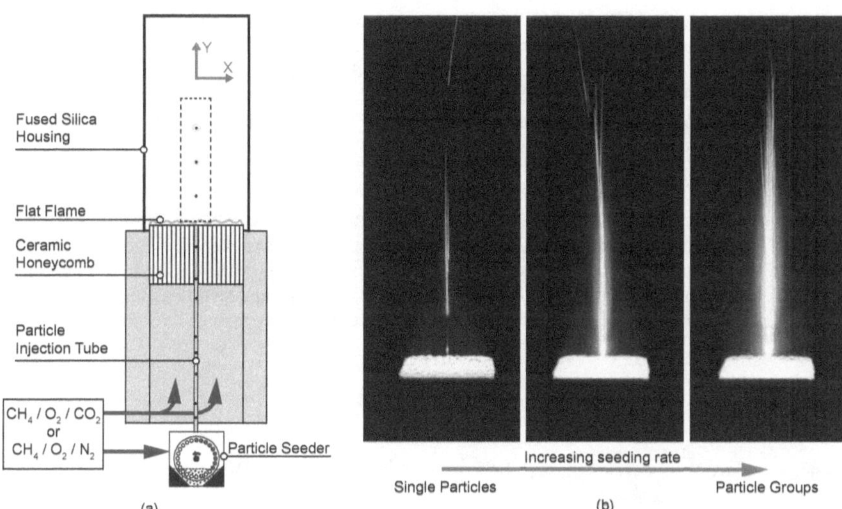

Figure 1: (a) Sketch of the used laminar flow reactor. (b) Luminosity images of single particles and particle clouds burning in the laminar flow reactor.

Colombian high-volatile bituminous (hvb) coal in a laminar flow reactor (LFR) in the high-speed laser laboratory of RSM institute at the Technical University of Darmstadt. Three-dimensional visualizations of the OH-rich zone in the vicinity of igniting coal particles are demonstrated with a laser scanning system consisting of a 10 kHz dye laser and an AOD. Simultaneous DBI is set up for in-situ detection of size, position, and number density of particles. In Section 2, the experimental methodology including the data processing procedures is introduced. In Section 3, the particle size distribution and the uncertainty of determination is analyzed. Then, the temporal and spatial evolution of OH-LIF signal distribution during the single particle volatile combustion is presented. The effect of particle size on the volatile flame stand-off distance determined at the maximum gradient of OH-LIF signals is analyzed. Finally, particle group combustion is evaluated based on three single-shot visualizations and the effect of PND on flame topology is discussed. This is the first data set of its kind for pulverized fuel. Based on the previously mentioned literature of experimental and numerical studies on pf ignition, these simultaneous measurements of OH radicals and PND provides novel data, which support the general understanding and detailed modelling of such flames.

2. Experimental Methods

2.1. Darmstadt laminar flow reactor

The experiments were performed in an enclosed LFR which is schematically illustrated in Fig. 1 (a). The reactor remained in the same configuration as reported in [27], which was mainly composed of a particle seeder with a central injection tube, a ceramic honeycomb ($\sim 8 \times 80$ mm^2), and a fused silica chimney enclosure. A fully premixed laminar methane flat flame was stabilized at 1.5 mm above the honeycomb surface. The inlet gas mixture was composed of N$_2$, O$_2$, and CH$_4$, which were issued at flow rates of 45.7, 18.1, and 5.2 L/min (at 0 °C and 1 bar). The exhaust gas composition contained 10 vol% oxygen and the gas velocity was about 1.6 m/s. The measured gas temperature was approximately 1800 K directly at the flame front downstream of the premixed flame, as reported in [27]. The particles were injected by a carrier gas flow through the central tube with inner diameter of 0.8 mm. The present experiments were carried out using an inert jet of N$_2$ at a flow rate of 0.03 L/min (at 0 °C and 1 bar), resulting in a slightly higher inlet velocity than that of the flat flame. The decrease in gas temperature along the jet center line was negligible [26]. As the particles entered the post-flame environment, they underwent a steep temperature gradient and thus a defined start of heating. The rectangular fused silica enclosure enabled free optical access from all sides of the LFR.

sieved to 90 to sizes of between 125 μm and 160 to 200 μm respectively. The proximate analysis of the coal composition was 3.5%m moisture(an), 36.9%m volatiles(wf), 54.4%m Cfix(wf) and 8.7%m ash(wf) [37]. Various particle seeding rates were achieved to investigate single particle and particle cloud combustion. Figure 1(b) illustrates the luminosity images of burning particle streaks with increasing particle seeding rates. The signals mainly result from the thermal radiation of soot and char particles. The images in Fig. 1(b) were recorded using a digital single-lens reflex (DSLR) camera with a constant exposure time of 1/8 second.

2.2. Simultaneous 3D OH-LIF and DBI measurements

The acousto-optic deflector being used for the presented experiments is schematically illustrated in Fig. 2. It resembles a typical AOD system, which consists of a function generator, a radio frequency (RF) driver, a piezo-transducer, and a crystal. Due to changes in the local crystal density (rarefaction and compression) caused by alternating DC voltages, the local index of refraction changes periodically. The crystal acts as an optical diffracting grating and is designed for Bragg diffraction. The interaction length L, namely the length the laser propagates through the crystal is sufficient if

$$L \geq \frac{\Lambda^2}{\lambda}, \tag{1}$$

where Λ denotes the acoustic wavelength and λ the optical wavelength. The Bragg angle θ_B is then defined as the function of three parameters, as shown in equation 2,

$$\theta_B \approx sin\theta_B = \frac{\lambda f}{2v}, \tag{2}$$

where v is the acoustic velocity within the crystal. If the incident angle $\theta_i = \theta_B$, the efficiency of the first diffraction order is at a maximum. The deflection angle θ_D, which is twice of the Bragg angle, can be derived as:

$$\theta_D = \frac{\lambda f}{v}. \tag{3}$$

Due to the fact that λ and v are constants restricted by the experimental hardware, θ_D is determined only by the acoustic frequency f. Figure 2(a) and (b) schematically show the minimum and maximum deflection angle, θ_{min} and θ_{max}, with acoustic frequencies f_{min} and f_{max}, respectively. By changing the acoustic frequency f, the scan angles or positions can be easily modulated, which is advantageous when compared to other scanners.

In this investigation, a water-cooled quartz crystal (D1340-aQ130, ISOMET) and a tuneable RF-driver (RFA3130, ISOMET) were employded. The control voltage ranging from 0‑10 V was generated from an arbitrary waveform generator (33500B Series, Keysight). The frequency bandwidth was 40 MHz, providing a maximum scan angle $\theta_{scan} = \theta_{max} - \theta_{min}$ of about 2 mrad. The quartz was coated for high transmission centered at the wavelength of 280 nm. The AOD was aligned for the strict Bragg condition at the center frequency f_c, resulting in a maximum diffraction efficiency of 70%. For all other frequencies/scan angles used, the diffraction efficiency was over 60%.

The experimental setup is illustrated in Fig. 3. A frequency-doubled dye laser (Credo, Sirah, Rohdamin 6G) was pumped by a diode-pumped solid-state laser (IS8II-E, EdgeWave) at 532 nm. The dye laser was tuned to 283.01 nm to excite the $Q_1(6)$ line of the A-X(1-0) transition of OH radicals. The laser was operated at 10 kHz repetition rate and produced 0.4 mJ pulse energy measured directly at the laser exit. The AOD scanner was operated with a step function for frequency modulation resulting in 10 individual scan angles at a 1 kHz scan frequency. More details regarding the synchronization of the AOD with the pulse laser are provided in [34]. The deflected laser beam was collected and parallelized by a cylindrical concave lens ($f_L = +2000$ mm) and focused at the burner center line. The parallelization was examined by traversing a beam profiling camera (WinCamD-LCM, Dataray) within ±20 mm from the burner center line along the laser path. The maximum scan angle of approximately 2 mrad and the focal length of 2000 mm jointly determine the scan depth of 3.8 mm along the z-axis. By employing a UV-coated cylindrical telescope composed of a plano-concave lens ($f_L = -100$ mm) and a plano-convex lens ($f_L = +500$ mm), the laser was expanded to a 20 mm high sheet with a thickness of 150 μm (FWHM).

The OH fluorescence signals were collected by a CMOS camera (HSS6, LaVision) coupled with a two stage in

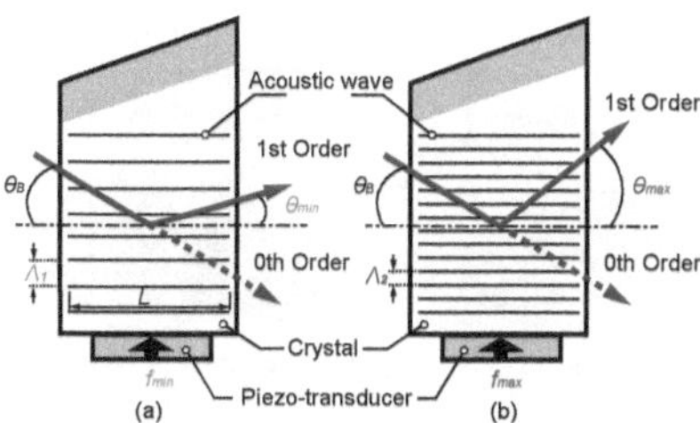

Figure 2: Typical acousto-optic deflector system and schematics of laser deflection in the Bragg regime with the incident beam with Bragg angle θ_B. The first order beam shows a deflection angle range between θ_{min} in (a) and θ_{max} in (b).

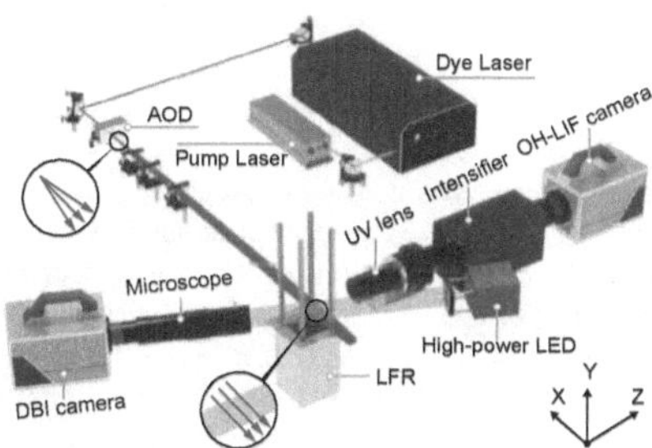

Figure 3: Experimental setup of simultaneous DBI and 3D OH-LIF measurements.

and the depth of field (DOF) of >5 mm was examined by imaging a calibration target at different positions along the z-axis. A band-pass filter (305 - 340 nm, Laser Components) with >50% transmission between 305 nm and 340 nm and the intensifier gate of 200 ns were applied to suppress broadband chemiluminescence and thermal radiation. A 2D field of view (FOV) of $18.7 \times 18.7 \text{ mm}^2$ and an in-plane pixel resolution of 24.4 µm were realized. The optical resolution of 100 µm was estimated by imaging and evaluating a Siemens star.

A DBI system was operated simultaneously to address the instantaneous particle number density. A high-power pulsed LED system (LPS, ILA) was angled by 17° with respect to the OH-LIF camera, as shown in Fig. 3. The LED was synchronized with the high-speed dye laser at 10 kHz with a pulse durations of 1 µs. A CMOS camera (v711, Phantom) equipped with a long-distance microscope (SK2, Infinity) was used to detect particle backlight images. To enhance the signal quality, a 525 nm band-pass filter with a bandwidth of 45 nm was employed, suppressing broadband radiation from soot and burning particles. The FOV of the DBI imaging was 11.8 (height) $\times$ 5.5 (width) mm^2 with a projected pixel resolution of 9.2 µm.

2.3. Data processing procedures

Prior to the 3D reconstruction, the OH-LIF images are corrected in terms of laser energy fluctuation and sheet inhomogeneity. The instantaneous laser profiles are extracted from the background OH-LIF signal with the assumption that the OH radicals, which result from dissociation reactions of water vapour in the exhaust gas, have a homogeneous distribution within the FOV. The temperature drop of the exhaust gas along the laser sheet height, which affects the OH-LIF signal level, is considered to be negligible [26]. Hence, the laser fluorescence signals depend only on the local laser energy in the chemical equilibrium. The raw signal is normalized based on the background intensity and is computed using 3×3 median and 3×3 Gaussian filters to reduce noise. Figure 4(a)-(c) exemplarily show three temporally sequential ($\Delta t = 200$ µs) pre-processed OH-LIF images of a single particle volatile flame at $z = -1.88, -1.04$, and -0.2 mm. At $z = -1.04$ mm, the intensity in the vicinity of the particle appears very low. A long dark streak in the negative side of the x-axis is observed, indicating that the particle is located approximately centrally within the laser

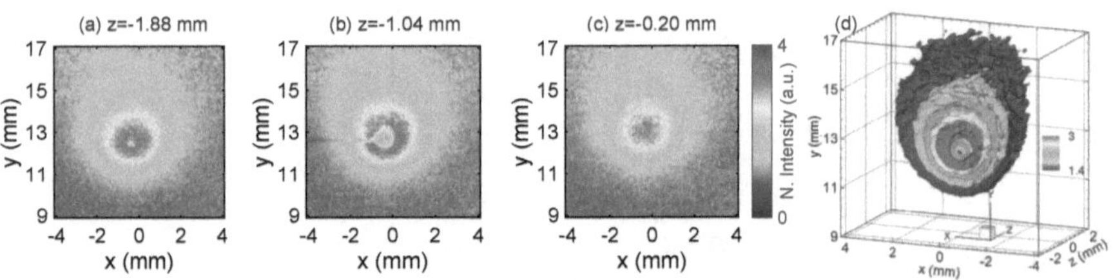

Figure 4: (a-c) Pre-processed OH-LIF images of a single-particle volatile flame at different positions of $z = -1.88\,\text{mm}$ $-1.04\,\text{mm}$, and $-0.2\,\text{mm}$. (d) 3D visualization of the reconstructed single-particle volatile flame with intensity iso-contours of OH-LIF signals.

and in front of the particle, respectively. A signal-to-noise ratio of approximately 13 is evaluated using the mean value and standard deviation of the intensity in the homogeneous region. The signal to background ratio varies between 5 and 10.

The z-discretization of the scanning laser sheets was not equidistant, which was a consequence of the non-linear response of the deflection angle to the modulation frequency of acoustic waves. The averaged z-discretization was 419 μm. For an accurate 3D reconstruction, precise positions of individual laser sheets were evaluated using the beam profiling camera. The raw images from each of the 10 planes were assigned to the corresponding z-positions to form a 3D signal matrix representing an entire detection volume of $18.7 \times 18.7 \times 3.8\,\text{mm}^3$. The volumetric matrix was linearly interpolated in the z-direction such that the spacing between the adjacent z planes was about 24.4 μm which was identical to the spatial pixel resolution in the x-y plane. The entire probe volume was then represented by $768 \times 768 \times 155$ voxels of $24.4\,\mu\text{m}^3$. Figure 4(d) shows a single-shot reconstruction of volumetric OH-LIF signals by using 10 planar images. The signals were filterd with a 3D Gaussian filter with the size of $5 \times 5 \times 5$ voxels to reduce spatial noise. The 3D volatile flame is represented with iso-contours of OH-LIF intensities. The slice crossing the particle center is displayed. Overall, the volatile combustion of the single particle reveals a symmetric structure, whereas the flame extends wider above the particle due to the slip velocity between the gas and particle. The signal drop in the vicinity of the coal particle and the dark streak observed in 2D images is well depicted in the 3D visualization.

Figure 5 (a) illustrates an instant DBI image of multiple coal particles. The DBI images cover the region from 2 mm above the burner surface up to 13.7 mm downstream. The pre-processing of DBI images included background subtraction and binarization using an adaptive threshold method [38]. The particle boundaries were detected based on the binary images. The equivalent diameter d_P was calculated with the assumption that the particle area enclosed by the boundary is equal to a circular area. Two individual coal particles with d_P of 126 μm and 172 μm are shown illustratively in the zoom-in views in Fig. 5(b) and (c), respectively. The detected boundaries provide a good approximation of the size of irregular particles.

3. Results and Discussions

3.1. Particle size and uncertainties

Figure 6 shows the probability density function (PDF) distributions of equivalent diameter d_P for coal particle samples and B, which were calculated separately. Approximately 70 particles were used for each sample and only single particles with an inter-particle distance L_x of $\geq 4\,\text{mm}$ were included. The particle-to-particle interaction affecting the local gas temperature and species distribution and the particle heating rate were excluded, thus guaranteeing the single-particle combustion mode. For each particle, only its first appearance in the FOV was used for calculating d_P. Both samples reveal a wider diameter distribution than the given specifications and both are slightly skewed to larger diameter values.

To examine the uncertainty of the diameter evaluation, a shadowgraph target was imaged and evaluated using the same algorithm. Figure 7 (a) shows a segment of discrete spot zones highlighted with dashed lines, which contain spots of in-prior defined diameters. The same algorithm for particle boundary detection was applied. The detected boundaries are plotted in red, and the equivalent diameter of each detected spot is labeled on the top-right corner. As expected, the

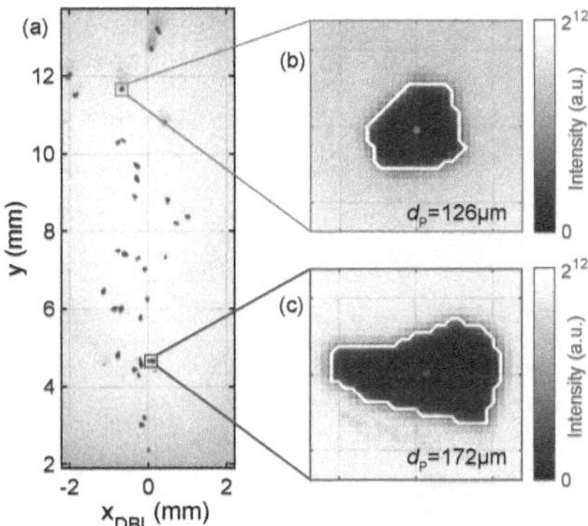

Figure 5: (a) A single-shot DBI image of multiple coal particles. (b-c) Two individual particles with zoom-in views with detected centroids and boundaries.

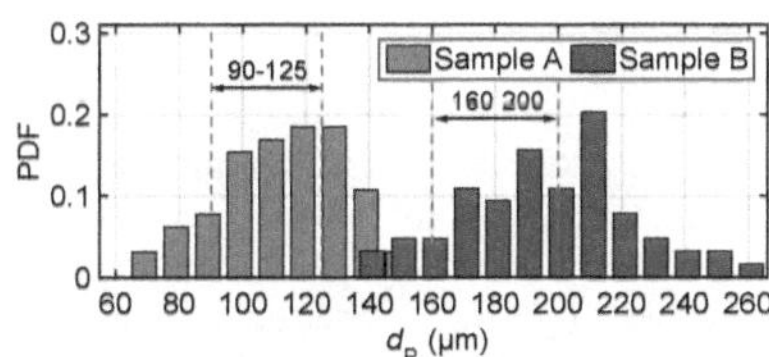

Figure 6: The PDF distribution of the particle equivalent diameter d_p of samples A and B.

to the insufficient local intensity gradients. By evaluating 500 segments, histograms of the detected spot diameter d is shown in Fig. 7(b). No valid peaks can be found in the vicinity of 40 μm. This can be explained by the significant overestimation of 20 μm spots, which leads to a PDF shifted towards larger diameters. Evident peaks of 57.5 μm, 97.5 μm and 200.5 μm are observed for 60 μm, 100 μm and 200 μm spots, respectively. For the spots between 60 and 200 μm, the error in the diameter determination is estimated to be smaller than 3.8%. Hence, it can be concluded that the particle diameter distribution in Fig. 6(a) is accurate within acceptable error margins. The wide d_P distribution of coal particles is probably due to the particle rotation and the irregular particle shape. The shift towards larger diameters can be explained by the swelling effect [39], which is influenced by the particle heating beginning from the flat flame front.

3.2. Volatile combustion of a single coal particle

In Fig. 8, the temporal evolution of a single-particle volatile flame is shown by means of an illustrative sequence of eight OH-LIF signal reconstructions, separated by 1 ms intervals. Slices crossing approximately the particle center are shown at each time step. The ignition time t_0 is defined at the scan sequence, when the first increase of signal intensity against the background level is observed. Because the signal at ignition appears very low and noisy, the selected sequence in Fig. 8 starts from $t_0 + 1$ ms, which is in the early stage of volatile combustion. The OH-LIF signal depends on the OH mole fraction Y_{OH}, temperature, quenching rate, and local laser intensity. The laser inhomogeneity and energy fluctuations was corrected shot-for-shot based on the laser beam profile and has no impact on the comparison. In the flame region, the OH-LIF signals respond non-linearly to the OH mole fraction. A quantitative estimation of Y_{OH} relies on the comparison with detailed numerical models taking non-linear effects into account, which still remains challenging for experiments. Mevel et al. conducted 1D simulations based on a three-level LIF model and extended to 2D simulation for applications in unsteady denotation combustion [40]. They concluded that, among other effects, the laser absorption is the dominant interference process in the LIF signal simulation. Boeck et al. [41] discussed

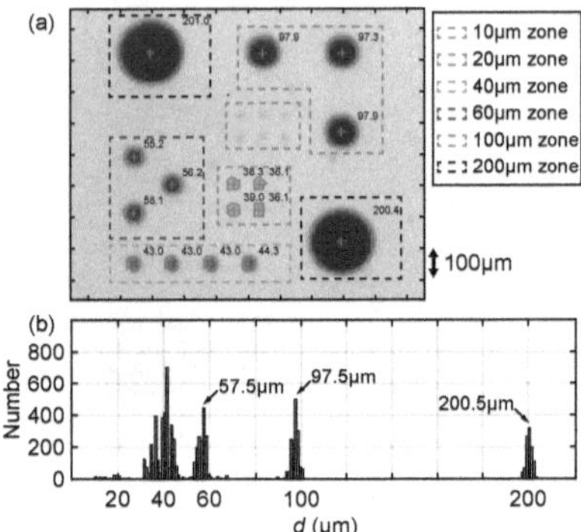

Figure 7: (a) A segment of the shadowgraph target with spot zones at discrete diameters. Equivalent diameters d are labeled in the top-right corner of each spot. (b) Histogram of the detected spot diameter d for the 6 spot sizes given in (a).

creased collisional quenching with the presence of extrem high OH number densities, which essentially impacted the LIF image quality. The complexity of local quenching, laser absorption and temperature effects can be considered in the prediction of Y_{OH} by including 1D simulation and non-linear LIF model, as recently demonstrated by Angelilli et al. [42]. Considering the low hydrogen content (generally lower than 5% by mass) in coal particles, the OH yield of volatile diffusion flames is expected to be far lower than that in H_2 gas flames in the mentioned studies. By examining the decrease of the background OH-LIF signals along the laser propagation direction, the laser energy absorption is estimated to be lower than 10% through the entire burner. In the present study, the actual OH number density is overestimated by its apparent LIF intensity, and it corresponds to the optically-thick regime reported in [43]. Nevertheless, for the determination of the flame stand-off distances, the non-linear effects of e.g. OH quenching and laser energy absorption are negligible. Dominated by local chemical reactions and temperatures, OH-LIF signals can be used as an effective marker of reaction zone for a qualitative comparison.

The first apparent OH-LIF signals are observed in downstream regions of the particles where higher temperatures of the oxidizer are expected. The ignition of the diffusion volatile flame begins within a flammable region of a proper mixture fraction and high enough temperatures. The location of igniton is affected by non-isotropic pyrolysis processes restricted by the non-uniform particle heating and the particle motion (slip and rotation) [27]. The downstream ignition is also dominated by the scalar dissipation rate which is relative low in the particle wake [44, 45]. The auto-ignition upstream of the particle may be prevented by the very high scalar dissipation rate. After the onset of ignition, the particle is enclosed by a spherical volatile flame which is possibly caused by a flame propagation towards upstream. In subsequent instants, the spatial inhomogeneity of OH-LIF signal distribution remains. In this stage, the devolatilization continues and can also be accelerated by the thermal radiation from exothermic reactions in the adjacent gas phase, which leads to a steeper rise of particle temperatures [44]. Moreover, a consistent increase of intensity and volume with sufficient OH-LIF signals is observed as the volatile combustion becomes more intensive. During the 6 ms depicted in Fig. 8, the peak OH-LIF intensity increases approximately threefold and the size of the signal volume approximately fivefold. In addition, the slight extension of the volatile flame above the particle as observed in 2D images is confirmed in all three dimensions. The elongated flame shape is considered to be a consequence of joint effects of the mixture fraction, the buoyancy of burnt gas, and the slip velocity. It is noteworthy that the full development of an enveloping volatile flame needs slightly longer than that discussed in the previous work [26], where a premixed carrier gas jet was used, mainly because the mixing of the inert jet with high temperature surrounding gas requires additional time.

To further study the temporal and spatial evolution of the volatile combustion, Figure 9 plots averaged radial intensity profiles for each instant from the onset of ignition. Only the left hemisphere of the flame is included. As previously mentioned, the signals at the onset of ignition are weak due to the low reactivity. In the first 3 milliseconds, no significant gradients along the radial axis are observed and the OH-LIF signals are in the same order of the background

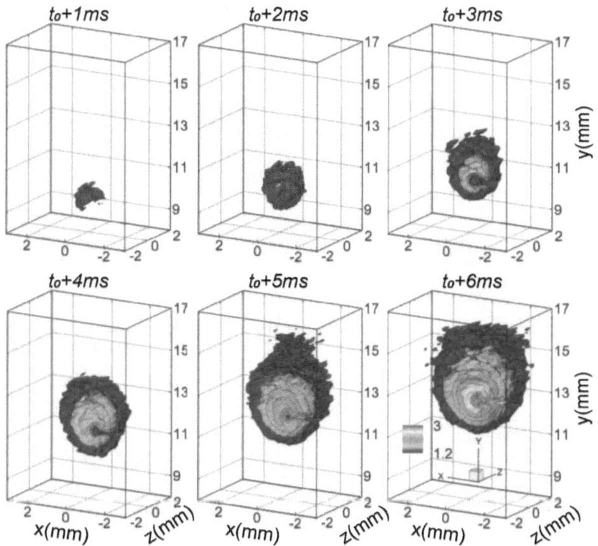

Figure 8: Temporal volumetric OH-LIF visualizations of a single particle volatile flame. The onset of ignition is denoted by t_0.

explained by the heat transfer from gas to solid phase and by a decrease of the gas temperature [46]. Thereafter, a significant rise of OH-LIF intensity is observed which is attributed to the accelerating reaction of the volatile diffusion flame. On the inner side of the reaction zone, steeper gradients are formed owing to a temperature rise towards the burnt gas region. Note that the intensity growth retards at 7 ms and stabilizes at 8 ms. The intensity nearest to the particle surface shows a consistent increase which can be explained by the initiating surface reactions with increasing particle temperatures (start of char combustion) [25]. Nevertheless, the effects of intensifier blurring and slight particle offset out of the laser sheet are also considered possible sources of bias. The distance between the OH-LIF peak and particle center remains approximately constant during the gas-phase reaction. This implies that the gas diffusion rate and devolatilization rate maintain roughly in balance. The location with steepest gradient, an indication on high reactivity, is contained within 0.3 mm from the particle center. The OH-LIF signals further decrease radially slowly approaching the background level where a sufficient OH mole fraction still exists due to the water dissociation reaction at high temperatures in chemical equilibrium.

3.3. Effect of the particle size

Figure 10 (a) shows the radial intensity I_{flame} of four individual volatile flames with increasing equivalent particle diameter d_{p} at the onset of ignition and 2 ms after the ignition. The selection of 2 ms guarantees that the volatile flame is still in the early stage of volatile combustion for differently sized particles. At the onset of ignition, I_{flame} slightly exceeds the background OH-LIF intensity and the radial profiles are similar for different particles. 2 ms after ignition, smaller particles exhibit higher maximum intensities $I_{\text{max, flame}}$. By evaluating $I_{\text{max, flame}}$ of all particles from samples A and B, Figure 10 (b) indicates a non-linear decrease of $I_{\text{max,flame}}$ with increasing d_{p}. The gas flame temperature probably drops with increasing d_{p} due to the higher energy transfer required from gas to solid phase for particle heating. Similar to the flame-wall interaction, the gas flame is weakened due to the heat loss caused by particle-flame interaction. Another possible reason for the decrease of $I_{\text{max,flame}}$ is the increased OH quenching rate due to the light hydrocarbons in the released volatiles. This effect should be more obvious for large particles with a high volatile mass. However, the volatile release rate is also influenced by the local gas temperature, and it is not clear which mechanism dominates without further knowledge on mixtures and temperatures associated with single particle flames. This raises new challenges for

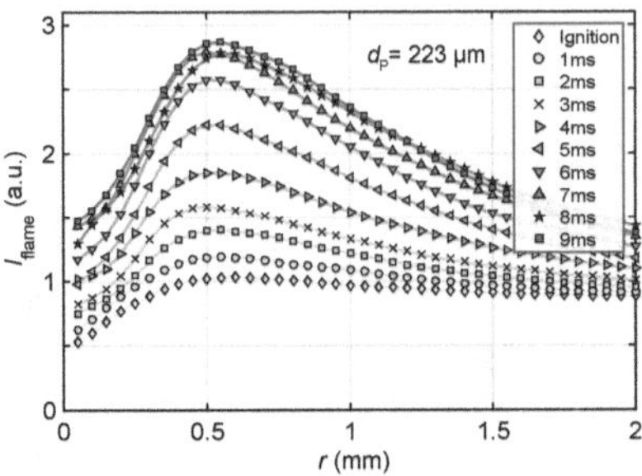

Figure 9: Spatially averaged radial OH-LIF intensity profiles of the volatile flame of a single coal particle.

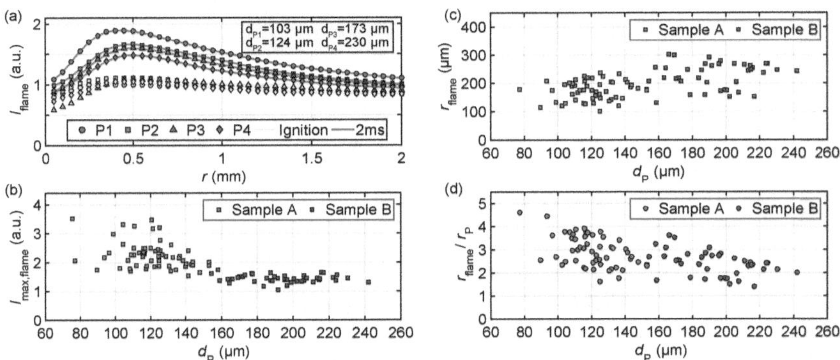

Figure 10: (a) Radial OH-LIF intensity profiles of four different single particle volatile flame at ignition and 2 ms after ignition. (b) Maximum flame OH-LIF intensity $I_{\mathrm{max,flame}}$, (c) flame stand-off distance r_{flame} and (d) dimensionless flame distance $r_{\mathrm{flame}}/r_{\mathrm{P}}$ over d_{P} at 2 ms after the onset of ignition.

calculated based on the maximum gradients of I_{flame}, which indicate the location of the peak heat release rate in gas flames. In Fig. 10 (c), r_{flame} has a slight tendency to increase with d_{P}. The entire volatile mass released from big particles is possibly higher than that from small particles, and thus push the flame front to the side of the oxidizer. However, the relative flame stand-off distance $r_{\mathrm{flame}}/r_{\mathrm{P}}$ in Fig. 10(d) reveals a contrary trend : the gas flame resides relatively closer to the particle center. As the initial gas temperature is equal for both size fractions, the large particles experience lower heating rates. As a consequence, the volatile release rate drops with increasing particle size which results in a flame being shifted closer to the particle surface.

The Fourier number is estimated to be about 0.02 ($Fo = \alpha\tau/d_{\mathrm{P}}^2$, where the thermal diffusivity α is estimated as $1 \times 10^{-7} m^{-2}/s$ for typical bituminous coal [47, 48]) for the particles with a diameter of 200 μm and a residence time of 8 ms. The residence time was considered as the ignition delay time referenced to the start of heating when particles cross the flame front. The Fourier number increases to 0.08 for a particle diameter of 100 μm and the same residence time. Although the volatile release is still a surface phenomenon as reported by [49], higher devolatilization rates still are expected for smaller particles due to enhanced thermal conduction. Note that the scatter in Fig. 10(b), (c), and (d) is mainly due to the inherent limitation of the line-of-sight measurements for d_{P}. The non-spherical particle shape and random orientation in three-dimensional space are alos dominant error sources. For a more accurate d_{P} determination, a tomographic reconstruction from different perspectives using multiple cameras and light sources is required, adding enormous complexity to the experimental investigation.

3.4. Effect of the particle number density

The effect of increasing particle number density on volatile combustion behaviour is investigated by simultaneous particle DBI imaging and 3D OH-LIF visualizations. In Fig. 11, three individual snapshots with different particle densities are compared. The DBI images providing the line-of-sight information about the particle number and location are shown on the left, and the 3D volatile flame structure represented by OH-LIF signals is on the right side. The FOV of DBI measurements is highlighted with dashed rectangles in the reconstructed flame structures.

In the low-density case of Fig. 11(a), particles show an enclosed volatile flame structure at the first glance. Upon closer inspection, nearly spherical flames can be observed in the near field of individual particles, with externally overlapping burnt gas regions, forming an apparent joint flame. The averaged particle distance of about 1 mm (PND $\sim 0.2 \times 10^9 \mathrm{m}^{-3}$) were determined from backlight imaging. Referring to the intensity profiles of the single particles previously mentioned, the main reaction zone indicated by the peak signal location is restricted within 0.5 mm from particle centers. It can therefore be concluded that the ignition process of individual particles is not strongly influenced by other particles at such a low number density. The volatile conversion in the near-field of particles proceeds individually, although interlinking of burnt gas regions exists in the outer periphery.

As the particle density increases, the particle-to-particle interaction becomes more visible. As shown in the DBI image of Fig. 11(b) (PND $\sim 0.5 \times 10^9 \mathrm{m}^{-3}$), the particles at $y \approx 9$ mm exhibit faint shadows in their surroundings indicating soot formation and the release of heavy volatiles [50]. A similar shadow is also observed for the low density case illustrated in Fig. 11(a), but at $y \approx 15$ mm towards the end of the volatile combustion phase. The earlier appearance of slight shadows at a higher PND can be explained by enhanced soot formation due to the local lack of oxygen available for fuel oxidation. The volatile diffusion flame is limited by the diffusivity of oxygen and volatile matter into each other. Taking a closer look at the OH-LIF signal distribution, no spherical flames are observed, implying extensive interaction between ignition and volatile flame of individual particles interact extensively. Additionally, there is a lack of OH-LIF signals in the center of the particle group measured indicating that the rich gas mixture is beyond the flammability limit leading to flame extinction. This argument is supported by the soot formation due to a local deficiency of oxygen. The gas temperature is also expected to be lower compared to that in the low density case due to the heat flux between gas and solid phase [51].

Increasing the particle number density further, the appearance of a lifted flame is observed without any high-temperature reactions forming OH radicals inside of the particle group, as shown in Fig. 11(c) (PND $\sim 1.2 \times 10^9 \mathrm{m}^{-3}$). The intensive release of heavy volatile matter and soot formation are observed in the backlight image. A globally fuel-rich mixture is expected in this case due to the massive volatile release of the particle cloud. The right side of the flame is barely detected due to laser absorption by the soot particles. In comparison with the middle particle density case, the non-flammable volume increases. Better flammability presumably exists at the outer layer of the particle cloud flame where oxygen is available.

Comparing the three individual visualizations, the volatile flame structure extends wider as the particle number density increases. As the PND increases, the ignition becomes more delayed mainly for two reasons. Firstly, the entire heat demand for particle heating increases, which results in a lower local gas temperature compared with single particle combustion. With increasing PND, the background signals (mainly resulting from water dissociation) surrounding particles decline dramatically prior to apparent ignition. This supports the argument of the higher energy demand needed for sufficient particle heating and accounts for the lower gas temperatures. Secondly, the particle heating rate depends on the temperature gradient between the gas and solid phase. Lower gas temperatures delay the particle temperature rise and thus the devolatilization process as well. This observation confirms the results of the numerical investigation with similar boundary conditions shown in [51]. From the diagnostics perspective, the 3D OH-LIF imaging enables the volumetric reconstruction of volatile flames in particle group combustion. As shown in Fig. 11, the OH-LIF signals from the gas phase often relate not only to the in-plane particles, but also the out-of-plane particles. While 2D images might contain incomplete or even misleading information when the flame has internal structures. The demonstrated potential benefits of the newly developed 3D laser scanning method make it advantageous for further experimental investigations.

The observations presented in this work are qualitatively in line with typical results of numerical simulations of pf particle streaks. Liu et al. report the the rise of the ignition delay time when the PND exceeds a critical particle size dependent value [10]. As they calculated the PND for the cold carrier flow, thermal expansion might somewhat bias the direct comparison to the present results. Additionally, they only report the sieve size distribution of the particles

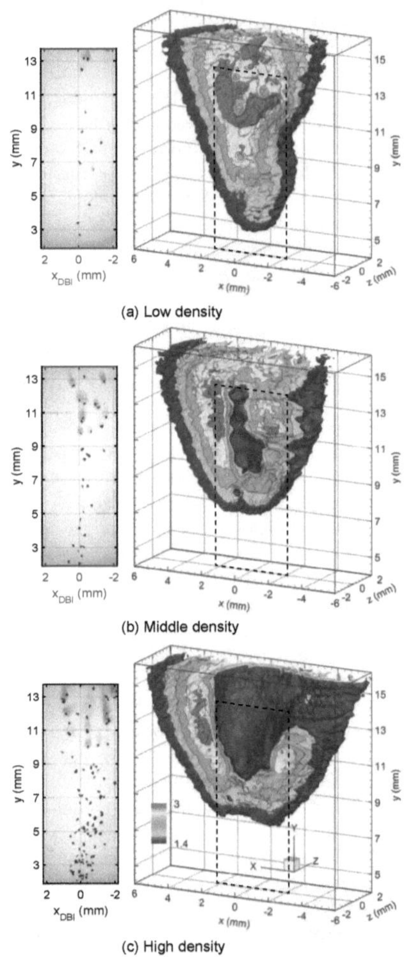

(a) Low density

(b) Middle density

(c) High density

Figure 11: Snapshots of volatile flame topology for three different particle densities. Simultaneous line-of-sight DBI images (left) with FOV highlighted by the rectangle in the 3D flame reconstruction (right).

without direct measurement of the particle size distribution. Liu et al also use a characteristic group number

$$G = 2\pi n R_c^2 d_P, \tag{4}$$

which is determined by the particle number in a cylindrical control volume ($n = $ PND), the column or cylinder radius R_c, and the particle diameter d_P. Further information on the calculation of group numbers can be found in [5]. The imaging approach in the present study allows for direct determination of all three parameters needed. The images of the particle streak in Fig. 11 show a streak radius which is in the range of $2 - 2.5 \times 10^{-3}$ m, when considering the outer particles. As the observed particle-to-particle interactions in the given case are observed for a PND around 0.5×10^9 m^{-3}, and particle radii by shadowgraphy are in the range of 150 µm, we observe the transition from single

more difficult to compare the results. As their reported changes in ignition delay times are most significant for group numbers at an equal range as the values reported in this work, the results seem to be comparable for roughly similar setups. It must be noted that setups with very different inflow geometries often report higher group numbers to cause faster ignition [52].

Prationo et al. [8] also reported a slight decrease in ignition delay times when the PND is increased up to 0.5 $\times 10^9$ m^{-3}, so even under thermal expansion in the radial direction of the particle streak, the values might be roughly comparable. Thermal expansion in the direction of the flow will cause a reduction of the PND, so it can be estimated that the PND in [8] is slightly lower than in the present study, nevertheless, the observed phenomena point in the same direction.

If the given PND of 0.5×10^9 m^{-3} is converted into a ratio of the particle distance L_x and particle diameter d_P, which many numerical studies use as a parameter, it is equivalent to $L_x/d_P = 10$ for $d_P = 125$ µm. Of course the experimental data is somewhat biased, as numerical studies often use fixed particle diameter, which is not the case in real particle samples. Tufano et al. [45, 53] investigated particle groups numerically. They extracted $L_x/d_P = 1 - 1000$ (smaller values indicate higher PND) from LES simulations of an existing furnace [54]. They found that at $L_x/d_P = 5$, very fuel-rich conditions in the inter-particle space occurred. At $L_x/d_P = 10$, they observed a significant reduction of this effect. As their simulations used a coal with similar volatile content (37%, compared to 36.9% present), an atmosphere with 21% O_2 and particles with $d_P = 100$ µm, the agreement between their simulations and the presented experiment is very good.

Farazi et al. studied particle group ignition [51], but with a few differences in the boundary conditions when compared with the present work. In their study, the atmosphere is dominated by CO_2 to study oxy-fuel combustion. They varied the relative velocity between gas and particle between 0.5 and 1 m/s, and investigated variations of the gas temperature between 1300 and 1500 K. A clear increase in ignition delay times is observed when the PND approaches 0.5×10^9 m^{-3}, and even half the particle concentration already initiates this effect. Although there is a phenomenological agreement between this work and their numerical results, further investigations with matching operational parameters will help to increase comparability and thus support the credibility of numerical simulations for plant simulations.

The direct linking between high-resolution OH-PLIF and LES/DNS simulations will be a suitable method to deepen the understanding of conversion and pollutant formation processes in coal combustion. The advantage of the presented technique lies in the level of detail, which describes the same length and time scales as these highly resolved simulations. Obviously, the reaction rates of gas phase are faster by several orders in magnitude. As the experimental results in the given work already show, that is possible to measure such factors with sufficient resolution, making a direct linkage between experiment and simulation is learly within reach.

4. Conclusion

The present work reported a recent experimental investigation on the volatile combustion of coal particles using simultaneous volumetric OH-LIF and diffuse backlight-illumination. An acousto-optic deflector was applied to sweep the laser beam of a dye laser system operating at 10 kHz. The three-dimensional reconstruction of the OH-LIF signal distribution was demonstrated within a volume of $\sim 18.7 \times 18.7 \times 3.8$ mm^3. The in-plane and out-of-plane spatial resolutions of the reconstruction were about 100 µm and 419 µm, respectively, while the temporal resolution was 1 ms. The DBI system allowed for the simultaneous detection of particle size and particle number density at 10 kHz. The ignition and volatile combustion of Colombian hvb coal particles with d_P ranging from $\sim 90 - 125$ µm and $\sim 160 - 200$ µm were investigated in a laminar flow reactor.

Several conclusions are summarized as follows:

1. The DBI technique is revealed as a high precision method for the determination of the particle equivalent diameter. The narrow error margins rely on the near-collimated light source, high spatial resolution of the camera and, a gradient-based boundary detection algorithm. While the uncertainty in the 2D case could be well characterized using a specified shadowgraph target, the 3D uneven shape of coal particles remains unresolved. Nevertheless, the high-resolved DBI measurements provided insights into the irregular particle shape and can be used to investigate particle rotation and swelling phenomena in future work.

2. The laser scanning OH-LIF measurement combined with the 3D reconstruction decodes the 3D volatile flame structures. A 6 ms sequence of reconstructed OH-LIF distributions illustratively visualizes the temporal and spa-

probably because of a sufficient amount of oxidizer and high temperatures. A slightly elongated envelope flame forms around single particles after ignition and propagates downstream with rapid rise of intensity and volume growth. The non-homogeneity of the OH distribution was considered to be a joint consequence of inhomogeneous devolatilization, buoyancy of the burnt gas, particle rotation, and slip velocity between particle and gas. For a better understanding of this multiphase phenomenon, high spatially resolved flow field measurements in the near-field of the particles are required.

3. The particle diameter has an essential influence on the flame structure and intensity for single particle combustion. The big particles seem to have a weaker volatile flame compared with smaller particles at the early combustion stage. With increasing d_P, the absolute flame stand-off distance r_{flame} increases, which is probably due to the increased volatile mass pushing the diffusion flame outwards. However, the inverse trend of the relative flame stand-off distance r_{flame}/r_P might indicate a lower specific volatile release rate (i.e. volatile mass per coal mass per second) associated with a larger particle volume. Thermal diffusion and mass diffusion in volatile flames, which could be depicted by measuring local gas temperatures and fuel mole fractions, need to be further investigated for the better understanding of the particle-flame interaction.

4. The particle number density has an essential influence on the topology of volatile flames. While at low densities only the burnt gas regions overlap, local flame extinction appears at higher particle densities. The transition of flame structures is a result of the decrease of local temperature and oxygen concentration. The insufficient local gas conditions promote the soot formation as the particle number density increases. To examine the predominant mechanism (e.g. quenching due to high heat loss or the fuel-rich mixture outside the flammability limits) for flame extinction, both experimental and numerical studies are needed. Nevertheless, the simultaneous particle number density and 3D flame structure measurements reported in this study present feasible and essential approaches for further experiments on the fundamentals of coal combustion.

5. From the diagnostic point of view, the potential benefit of the presented techniques in turbulent flames is of great interest for combustion research. While the DBI measurements can easily be extended for high temporal and spatial resolution, laser scanning is limited by the laser repetition rates, which are commonly limited to 10 kHz for commercial high-speed lasers. As a first attempt, the capability and feasibly of the 3D imaging with AOD has been demonstrated in different non-reactive and reactive turbulent flow [34, 35]. For individual applications in turbulent conditions, a good compromise between spatial (i.e. number of scan planes) and temporal resolution (i.e. scan frequency) needs to be carefully considered.

Declarations of Interest

Martin Schiemann and Benjamin Böhm are co-authors on this paper and as the Guest Editor of this VSI Clean Fuel Conversion in RSER. They were blinded to this paper during the review process, and the paper was independently handled by Aoife Mary M Foley.

References

[1] V. Masson-Delmotte, H.-O. Pörtner, J. Skea, A. Pirani, R. Pidcock, Y. Chen, Global warming of 1.5°C, Special report, IPCC, [Geneva, Switzerland], 2018.

[2] D. Stolten, Efficient carbon capture for coal power plants, Chemical Engineering & Technology 35 (3) (2012) 407. `doi:10.1002/ceat.201290011`.

[3] R. H. Essenhigh, M. K. Misra, D. W. Shaw, Ignition of coal particles: A review, Combustion and Flame 77 (1) (1989) 3–30. `doi:10.1016/0010-2180(89)90101-6`.

[4] K. Annamalai, W. Ryan, Interactive processes in gasification and combustion—II. isolated carbon, coal and porous char particles, Progress in

[5] K. Annamalai, W. Ryan, S. Dhanapalan, Interactive processes in gasification and combustion—Part III: Coal/char particle arrays, streams and clouds, Progress in Energy and Combustion Science 20 (6) (1994) 487–618. doi:10.1016/0360-1285(94)90002-7.

[6] A. C. Sarroza, T. D. Bennet, C. Eastwick, H. Liu, Characterising pulverised fuel ignition in a visual drop tube furnace by use of a high-speed imaging technique, Fuel Processing Technology 157 (2017) 1–11. doi:10.1016/j.fuproc.2016.11.002.

[7] Y. Yuan, S. Li, F. Zhao, Q. Yao, M. B. Long, Characterization on hetero-homogeneous ignition of pulverized coal particle streams using ch* chemiluminescence and 3 color pyrometry, Fuel 184 (2016) 1000–1006. doi:10.1016/j.fuel.2015.11.032.

[8] W. Prationo, L. Zhang, Influence of steam on ignition of victorian brown coal particle stream in oxy-fuel combustion: In - situ diagnosis and transient ignition modelling, Fuel 181 (2016) 1203–1213. doi:10.1016/j.fuel.2016.03.003.

[9] Z. Zeng, T. Zhang, S. Zheng, W. Wu, Y. Zhou, Ignition and combustion characteristics of coal particles under high-temperature and low-oxygen environments mimicking MILD oxy-coal combustion conditions, Fuel 253 (2019) 1104–1113. doi:10.1016/j.fuel.2019.05.101.

[10] Y. Liu, M. Geier, A. Molina, C. R. Shaddix, Pulverized coal stream ignition delay under conventional and oxy-fuel combustion conditions, International Journal of Greenhouse Gas Control 5 (2011) S36–S46. doi:10.1016/j.ijggc.2011.05.028.

[11] Y. Xu, S. Li, Q. Yao, Y. Yuan, Investigation of steam effect on ignition of dispersed coal particles in O_2/N_2 and O_2/CO_2 ambiences, Fuel 233 (2018) 388–395. doi:10.1016/j.fuel.2018.06.047.

[12] L. Chen, S. Z. Yong, A. F. Ghoniem, Oxy-fuel combustion of pulverized coal: Characterization, fundamentals, stabilization and CFD modeling, Progress in Energy and Combustion Science 38 (2) (2012) 156–214. doi:10.1016/j.pecs.2011.09.003.

[13] S. Seepana, S. Jayanti, Steam-moderated oxy-fuel combustion, Energy Conversion and Management 51 (10) (2010) 1981–1988. doi:10.1016/j.enconman.2010.02.031.

[14] A. Adeosun, Z. Xiao, Z. Yang, Q. Yao, R. L. Axelbaum, The effects of particle size and reducing-to-oxidizing environment on coal stream ignition, Combustion and Flame (2018). doi:10.1016/j.combustflame.2018.05.003.

[15] Y. Yuan, S. Li, Y. Xu, Q. Yao, Experimental and theoretical analyses on ignition and surface temperature of dispersed coal particles in O_2/N_2 and O_2/CO_2 ambients, Fuel 201 (2017) 93–98. doi:10.1016/j.fuel.2016.09.079.

[16] Y. Yuan, S. Li, G. Li, N. Wu, Q. Yao, The transition of heterogeneous–homogeneous ignitions of dispersed coal particle streams, Combustion and Flame 161 (9) (2014) 2458–2468. doi:10.1016/j.combustflame.2014.03.008.

[17] F. Tabet, I. Gökalp, Review on CFD based models for co-firing coal and biomass, Renewable and Sustainable Energy Reviews 51 (2015) 1101–1114. doi:10.1016/j.rser.2015.07.045.

[18] A. Ramos, E. Monteiro, A. Rouboa, Numerical approaches and comprehensive models for gasification process: A review, Renewable and Sustainable Energy Reviews 110 (2019) 188–206. doi:10.1016/j.rser.2019.04.048.

[19] R. Knappstein, G. Kuenne, A. Ketelheun, J. Köser, L. Becker, S. Heuer, M. Schiemann, V. Scherer, A. Dreizler, A. Sadiki, J. Janicka, Devolatilization and volatiles reaction of individual coal particles in the context of FGM tabulated chemistry, Combustion and Flame 169 (2016) 72–84. doi:10.1016/j.combustflame.2016.04.014.

[20] T. Sayadi, S. Farazi, S. Kang, H. Pitsch, Transient multiple particle simulations of char particle combustion, Fuel 199 (2017) 289–298. doi:10.1016/j.fuel.2017.02.096.

[21] S. Farazi, A. Attili, S. Kang, H. Pitsch, Numerical study of coal particle ignition in air and oxy-atmosphere, Proceedings of the Combustion Institute 37 (3) (2019) 2867–2874. doi:10.1016/j.proci.2018.07.002.

[22] D. M. Grant, R. J. Pugmire, T. H. Fletcher, A. R. Kerstein, Chemical model of coal devolatilization using percolation lattice statistics, Energy & Fuels 3 (2) (1989) 175–186. doi:10.1021/ef00014a011.

[23] K. Xu, Y. Wu, Z. Wang, Y. Yang, H. Zhang, Experimental study on ignition behavior of pulverized coal particle clouds in a turbulent jet, Fuel 167 (2016) 218–225. doi:10.1016/j.fuel.2015.11.027.

[24] Y. Xu, S. Li, Q. Gao, Q. Yao, J. Liu, Characterization on ignition and volatile combustion of dispersed coal particle streams: In situ diagnostics and transient modeling, Energy & Fuels 32 (9) (2018) 9850–9858. doi:10.1021/acs.energyfuels.8b01322.

[25] J. Köser, L. G. Becker, N. Vorobiev, M. Schiemann, V. Scherer, B. Böhm, A. Dreizler, Characterization of single coal particle combustion within oxygen-enriched environments using high-speed OH-PLIF, Applied Physics B 121 (4) (2015) 459–464. doi:10.1007/s00340-015-6253-3.

[26] J. Köser, L. G. Becker, A.-K. Goßmann, B. Böhm, A. Dreizler, Investigation of ignition and volatile combustion of single coal particles within oxygen-enriched atmospheres using high-speed OH-PLIF, Proceedings of the Combustion Institute 36 (2) (2017) 2103–2111. doi:10.1016/j.proci.2016.07.083.

[27] J. Köser, T. Li, N. Vorobiev, A. Dreizler, M. Schiemann, B. Böhm, Multi-parameter diagnostics for high-resolution in-situ measurements of single coal particle combustion, Proceedings of the Combustion Institute 37 (3) (2019) 2893–2900. doi:10.1016/j.proci.2018.05.116.

[28] I. Boxx, C. Heeger, R. Gordon, B. Böhm, M. Aigner, A. Dreizler, W. Meier, Simultaneous three-component PIV/OH-PLIF measurements of a turbulent lifted, C_3H_8-argon jet diffusion flame at 1.5khz repetition rate, Proceedings of the Combustion Institute 32 (1) (2009) 905–912. doi:10.1016/j.proci.2008.06.023.

[29] R. Wellander, M. Richter, M. Aldén, Time resolved, 3D imaging (4D) of two phase flow at a repetition rate of 1 kHz, Optics express 19 (22) (2011) 21508–21514. doi:10.1364/OE.19.021508.

[30] V. A. Miller, V. A. Troutman, R. K. Hanson, Near-kHz 3D tracer-based LIF imaging of a co-flow jet using toluene, Measurement Science and Technology 25 (7) (2014) 75403. doi:10.1088/0957-0233/25/7/075403.

[31] J. Olofsson, M. Richter, M. Aldén, M. Augé, Development of high temporally and spatially (three-dimensional) resolved formaldehyde measurements in combustion environments, Review of Scientific Instruments 77 (1) (2006) 13104. doi:10.1063/1.2165569.

[32] J. Weinkauff, M. Greifenstein, A. Dreizler, B. Böhm, Time resolved three-dimensional flamebase imaging of a lifted jet flame by laser scanning, Measurement Science and Technology 26 (10) (2015) 105201. doi:10.1088/0957-0233/26/10/105201.

[33] G. Römer, P. Bechtold, Electro-optic and acousto-optic laser beam scanners, Physics Procedia 56 (2014) 29–39. doi:10.1016/j.phpro.2014.08.092.

Applied Physics B 123 (3) (2017) 1243. doi:10.1007/s00340-017-6663-5.

[35] T. Li, B. Zhou, F. Jonathan, A. Dreizler, B. Böhm, High-speed volumetric imaging of formaldehyde in a lifted turbulent jet flame using an acousto-optic deflector, Experiments in Fluids 61 (4) (2020) 2903. doi:10.1007/s00348-020-2915-y.

[36] J. Bode, J. Schorr, C. Krüger, A. Dreizler, B. Böhm, Influence of the in-cylinder flow on cycle-to-cycle variations in lean combustion DISI engines measured by high-speed scanning-PIV, Proceedings of the Combustion Institute 37 (4) (2019) 4929–4936. doi:10.1016/j.proci.2018.07.021.

[37] N. Vorobiev, M. Geier, M. Schiemann, V. Scherer, Experimentation for char combustion kinetics measurements: Bias from char preparation, Fuel Processing Technology 151 (2016) 155–165. doi:10.1016/j.fuproc.2016.05.005.

[38] J. Pareja, A. Johchi, T. Li, A. Dreizler, B. Böhm, A study of the spatial and temporal evolution of auto-ignition kernels using time-resolved tomographic OH-LIF, Proceedings of the Combustion Institute 37 (2) (2019) 1321–1328. doi:10.1016/j.proci.2018.06.028.

[39] D. Zeng, M. Clark, T. Gunderson, W. C. Hecker, T. H. Fletcher, Swelling properties and intrinsic reactivities of coal chars produced at elevated pressures and high heating rates, Proceedings of the Combustion Institute 30 (2) (2005) 2213–2221. doi:10.1016/j.proci.2004.07.038.

[40] R. Mével, D. Davidenko, J. M. Austin, F. Pintgen, J. E. Shepherd, Application of a laser induced fluorescence model to the numerical simulation of detonation waves in hydrogen–oxygen–diluent mixtures, International Journal of Hydrogen Energy 39 (11) (2014) 6044–6060. doi:10.1016/j.ijhydene.2014.01.182.

[41] L. R. Boeck, R. Mével, T. Fiala, J. Hasslberger, T. Sattelmayer, High-speed OH-PLIF imaging of deflagration-to-detonation transition in H₂–air mixtures, Experiments in Fluids 57 (6) (2016) 133. doi:10.1007/s00348-016-2191-z.

[42] L. Angelilli, P. P. Ciottoli, R. Malpica Galassi, T. Guiberti, W. Boyette, F. Hernandez Perez, G. Magnotti, W. L. Roberts, M. Valorani, H. G. Im, A new oh fluorescence signal-to-OH mole fraction conversion model formulation and calibration, in: AIAA Scitech 2020 Forum, American Institute of Aeronautics and Astronautics, Reston, Virginia, 01062020, p. 1. doi:10.2514/6.2020-1279.

[43] R. Mével, Optical regime diagram of the shock tube/pulsed laser-induced fluorescence imaging technique, Chemical Physics Letters 730 (2019) 283–288. doi:10.1016/j.cplett.2019.06.009.

[44] M. Vascellari, H. Xu, C. Hasse, Flamelet modeling of coal particle ignition, Proceedings of the Combustion Institute 34 (2) (2013) 2445–2452. doi:10.1016/j.proci.2012.06.152.

[45] G. L. Tufano, O. T. Stein, B. Wang, A. Kronenburg, M. Rieth, A. M. Kempf, Coal particle volatile combustion and flame interaction. Part I: Characterization of transient and group effects, Fuel 229 (2018) 262–269. doi:10.1016/j.fuel.2018.02.105.

[46] R. Knappstein, G. Kuenne, T. Meier, A. Sadiki, J. Janicka, Evaluation of coal particle volatiles reaction by using detailed kinetics and fgm tabulated chemistry, Fuel 201 (2017) 39–52. doi:10.1016/j.fuel.2016.10.033.

[47] J. Deng, Q.-W. Li, Y. Xiao, C.-M. Shu, Experimental study on the thermal properties of coal during pyrolysis, oxidation, and re-oxidation, Applied Thermal Engineering 110 (2017) 1137–1152. doi:10.1016/j.applthermaleng.2016.09.009.

[48] H. Wen, J.-h. Lu, Y. Xiao, J. Deng, Temperature dependence of thermal conductivity, diffusion and specific heat capacity for coal and rocks from coalfield, Thermochimica Acta 619 (2015) 41–47. doi:10.1016/j.tca.2015.09.018.

[49] R. Rajasegar, D. C. Kyritsis, Experimental investigation of coal combustion in coal-laden methane jets, Journal of Energy Engineering 141 (2) (2015). doi:10.1061/(ASCE)EY.1943-7897.0000228.

[50] T. H. Fletcher, J. Ma, J. Rigby, A. Brown, B. Webb, Soot in coal combustion systems, Prog. Energy Combustion Sci. 23 (1997) 283–301.

[51] S. Farazi, J. Hinrichs, M. Davidovic, T. Falkenstein, M. Bode, S. Kang, A. Attili, H. Pitsch, Numerical investigation of coal particle stream ignition in oxy-atmosphere, Fuel 241 (2019) 477–487. doi:10.1016/j.fuel.2018.11.108.

[52] W. Ryan, K. Annamalai, Group ignition of a cloud of coal particles, Journal of Heat Transfer 113 (3) (1991) 677–687. doi:10.1115/1.2910618.

[53] G. L. Tufano, O. T. Stein, B. Wang, A. Kronenburg, M. Rieth, A. M. Kempf, Coal particle volatile combustion and flame interaction. Part II: Effects of particle reynolds number and turbulence, Fuel 234 (2018) 723–731. doi:10.1016/j.fuel.2018.07.054.

[54] M. Rieth, F. Proch, M. Rabaçal, B. M. Franchetti, F. Cavallo Marincola, A. M. Kempf, Flamelet LES of a semi-industrial pulverized coal furnace, Combustion and Flame 173 (2016) 39–56. doi:10.1016/j.combustflame.2016.07.013.

Appendix F (Paper VI):
Volumetric measurements of group particle combustion

Investigation of the transition from single to group coal particle combustion using high-speed scanning OH-LIF and diffuse backlight-illumination

Tao Li[a,*], Christopher Geschwindner[a], Jan Köser[a], Martin Schiemann[b], Andreas Dreizler[a], Benjamin Böhm[a]

[a]*Reaktive Strömungen und Messtechnik, Technische Universität Darmstadt, Otto-Berndt-Straße 3, 64287 Darmstadt, Germany*
[b]*Lehrstuhl für Energieanlagen und Energieprozesstechnik, Ruhr-Universität Bochum, Universitätsstraße 150, 44801 Bochum, Germany*

*Corresponding author: Tao Li
Email address:* `tao.li@rsm.tu-darmstadt.de` (Tao Li)

Preprint submitted to Proceedings of the Combustion Institute

May 6, 2020

1. Introduction

Flame stability in pulverised fuel (PF) flames depends on the ignition characteristics of the particle cloud. In PF particle groups, heat and mass transfer effects in the inter-particle space play an important role and cause the ignition process to differ from single particle ignition [1, 2]. Experimental investigations on single particles [3, 4, 5, 6] and particle groups [7, 8] have been carried out with different optical methods . Laser-induced fluorescence of OH (OH-LIF) was employed on single coal particles to resolve the enveloping volatile flame surrounding high-volatile bituminous (hvb) coal particles with characteristic diameters of $100\,\mu m$ in time and space [5]. Combinations of OH-LIF with backlight-illumination and luminescence measurements provided correlated statistics between ignition time, volatile combustion duration and particle size for a hvb coal [4]. Other studies investigated CH^* chemiluminescence in correlation to alkali release for single biomass particles [9].

With numerical modeling reaching higher spatial and temporal resolutions, different groups [10, 11, 12] have investigated ignition on the single particle scale, e.g. [11] comparing their results on single particle ignition to the experiments reported in [5]. Numerical studies on coal particle group ignition with a resolution of or below the particle length-scale were carried out by different groups. Farazi et al. investigated a stream of coal particles under oxy-fuel conditions varying the particle number density (PND) [13]. Starting at low PND, they observed a decrease in ignition delay time with increasing particle concentration, before smaller inter-particle distances lengthened ignition delay time. Tufano et al. observed a transition from individual particle combustion to group combustion when reducing the inter-particle distance from 30 to 10 particle diameters (d_P) [14].

This work demonstrates the application of OH-LIF to scan the volatile flame of igniting coal particle groups in combination with diffuse backlight-illumination (DBI) of particles to correlate the structural characteristics of the volatile flame to PND. The experiments are carried out in a laminar premixed burner and provide general understanding and quantitative results on transient particle group combustion.

2. Experimental methodology

2.1. Laminar flow reactor

The experiment was performed in an enclosed laminar flow reactor (LFR, Fig. 1), which was already investigated in previous work [4, 5]. A laminar premixed flat flame (FF) was stabilized approximately 1.8 mm above a ceramic honeycomb surface. With an inlet gas mixture of N_2, O_2 and CH_4, the exhaust gas contained 10 vol% oxygen with a velocity of approximately 1.5 m/s. The maximum gas phase temperature of 1800 K was measured using nanosecond rotational-vibrational coherent anti-Stokes Raman spectroscopy (CARS) [4]. A minor temperature decrease of approximately 1 K/mm downstream along the y-axis was noticed. Coal particles were injected using an inert N_2 carrier flow through a central tube (Coal) with an inner diameter of 0.8 mm. The bulk velocity of the carrier gas was slightly increased to carry a large number of particles. The heat-up point was well-defined by particles crossing the enclosed premixed flame front and the steep rise of the gas temperature. The maximum particle heating rate in this configuration was estimated in the order of 10^5 K/s. The quadratic quartz glass enclosure enabled wide optical access from all sides of the LFR.

The coal investigated in this study was a hvb Colombian coal with a mean diameter of 120 μm. The proximate analysis of the coal composition was 3.5 %$_m$ moisture(an), 36.9 %$_m$ volatiles(wf), 54.4 %$_m$ C_{fix}(wf) and 8.7 %$_m$ ash(wf).

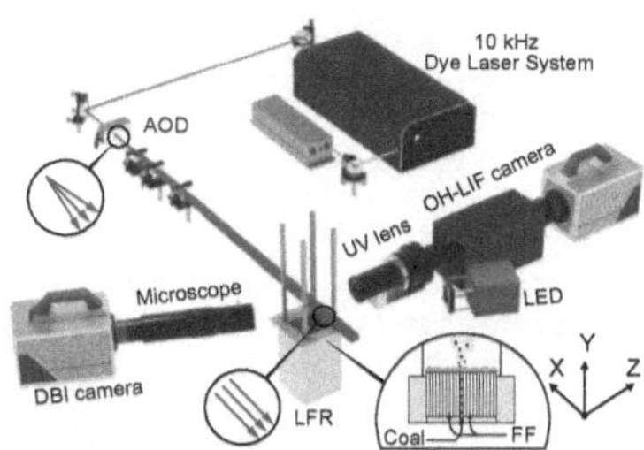

Figure 1: Experimental setup of simultaneous volumetric OH-LIF and DBI measurements. Zoomed-in views schematically highlight the laser path and the ceramic honeycomb.

2.2. Scanning OH-LIF

Figure 1 illustrates the experimental setup and arrangement of the diagnostic systems. A frequency-doubled dye laser (Credo, Sirah, Rhodamin 6G) was pumped by a diode-pumped solid-state laser (IS8II-E, EdgeWave) at 532 nm. The dye laser was tuned to 283.01 nm to excite the $Q_1(6)$ line of the A-X(1-0) transition system of the OH-radical. The dye laser system was operated at 10 kHz repetition rate and produced a pulse energy of 0.4 mJ measured directly at the exit. A water-cooled AOD (D1340, ISOMET) was employed to sweep the laser beam at 1 kHz resulting in 10 individual scan angles per sequence. The particle shift within one scan sequence is estimated to 500 μm. The crystal was coated for high UV transmission and had an overall efficiency of 70 %. For more details of laser scanning refer to [15, 16]. The deflected beams were parallelized and focused at the LFR centerline. The laser beam was expanded to a 20 mm high sheet with a thickness of 150 μm (FWHM). The parallelization of laser sheets was examined by traversing a beam profiling camera (WinCamD-LCM, Dataray) ± 20 mm from the measurement point along the laser path. The entire scan depth was 3.8 mm and had a standard deviation of <100 μm.

The OH fluorescence signal was collected by a CMOS camera (HSS6, LaVision) coupled with a two-stage intensifier (HS-IRO, LaVision). The camera was equipped with a UV-achromatic lens (Halle, $f = 150$ mm, $f/2.5$) and the depth of field of >5 mm was examined by imaging a calibration target at different positions along the z-axis. A band-pass filter $(300 \pm 15$ nm) and an intensifier gate of 200 ns were applied to suppress broadband chemiluminescence and thermal radiation. The field of view (FOV) was 18.7×18.7 mm^2 with a projected pixel size of 24.4 μm. The in-plane resolution was evaluated to be approximately 100 μm using a Siemens star. The out-of-plane resolution of 420 μm was restricted to the distance of adjacent laser planes.

2.3. High-speed DBI

A high-speed DBI imaging system was operated simultaneously to measure the in-situ PND. A high-power pulsed LED (IPS, ILA) was inclined by 17° to the OH-LIF camera and operated at 10 kHz with a peak wavelength of 525 nm and a pulse duration of 1 μs. A CMOS camera (v711, Phantom) equipped with a long-distance microscope (SK2, Infinity) was used to detect the particle backlight images. To enhance the signal quality, a band-pass filter $(525 \pm 22.5$ nm) was employed, suppressing broadband radiation from soot and particle luminosity. The FOV was 11.8×5.5 mm^2 with a projected pixel resolution of 9.2 μm. The region from 1.9 to 13.7 mm along the y-axis above the LFR surface was imaged.

2.4. Volumetric reconstruction of OH-LIF signals

The volumetric reconstrucion of the OH-LIF signal was performed within an effective detection volume of $18.7 \times 18.7 \times 3.8$ mm^3, capturing the entire region from the LFR surface to the region in which particles ignited and the volatile flame developed. The 2D images were corrected in terms of laser energy fluctuations and in-sheet inhomogeneity. The instantaneous laser profiles were extracted from the background OH-LIF signal, which resulted from water vapor dissociation in the partially homogeneous exhaust gas of the premixed flat flame. The gas temperature drop in downstream direction was considered to be negligible. The 2D OH-LIF signals (I_{OH}) were normalized with respect to the background OH-LIF intensity. A sequence of OH-LIF images of a single-particle volatile flame is illustrated in Fig. 2. The signals increase dramatically within several milliseconds after the ignition. The dark streak on the left-hand side indicates that the particle locates in the laser sheet in these instances. The signal-to-noise ratio of >10 is estimated by calculating the mean and the standard deviation of OH-LIF intensities in the background. The 3D intensity matrix formed from 10 individual 2D images of each scan sequence was linearly interpolated along the z-axis with the same spacing as the pixel resolution. Three-dimensional Gaussian and median filters, both with a size of $5 \times 5 \times 5$ voxels, were applied for noise reduction.

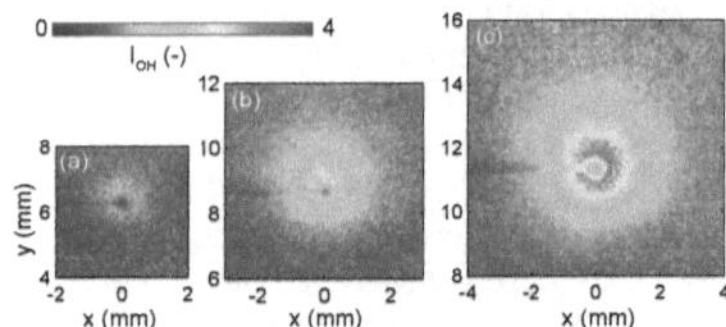

Figure 2: A temporal sequence of 2D OH-LIF images of a single-particle volatile flame at (a) 1 ms, (b) 3 ms and (c) 5 ms after the onset of ignition with the same image scaling applied.

The OH-LIF reconstruction is first demonstrated on single particle combustion. When a particle is exposed to the hot ambient gas, heat transfer from the gaseous to the solid phase heats the particle. With the increasing devolatilization rate, the particle releases volatiles which diffuse into the surrounding gas. The homogeneous ignition is defined here at the instant when the OH-LIF intensity increases for the first time above the background level. Figures 3a-c exemplarily show reconstructed OH-LIF intensity iso-contours of a single particle volatile flame at 1, 2 and 3 ms after ignition. The coordinate system (x, y, z) is centered at the particle jet exit. Even at an early stage of volatile combustion, an enclosed spherical flame forms over the circumference and shows temporal growth in size and intensity. This indicates that the homogeneous reaction rate is greater than the heat loss rate to the adjacent region. The OH-

3

LIF iso-surfaces show a slightly elongated flame structure with a maximum intensity downstream of the particle due to the slip velocity between particle and surrounding gas. The high OH concentration and gas temperature were also observed in other experimental [3, 4] and numerical investigations [12, 13, 14].

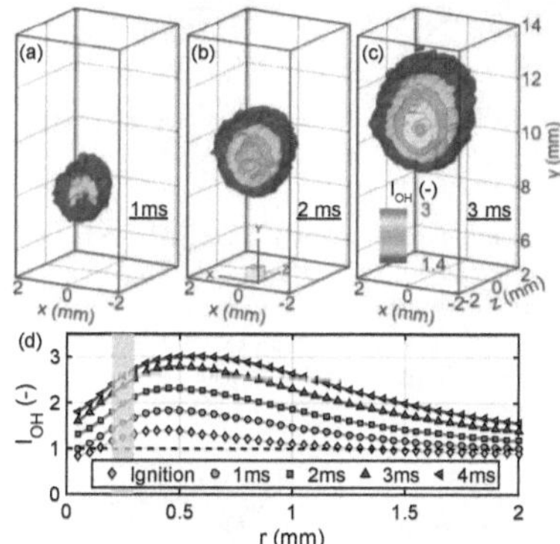

Figure 3: (a) - (c) Instantaneous OH-LIF reconstruction of a single particle flame at 1, 2 and 3 ms after ignition. (d) Mean radial intensity profiles starting from the onset of ignition.

In Fig. 3d, the average radial OH-LIF intensity profile is evaluated for 4 ms starting at the onset of ignition, only including the left semi-spherical flame to avoid bias from shadowing by particles. The intensity rises steeply after ignition to triple the background intensity. The location of the reaction zone is determined by the maximum OH gradients. It remains stable at a distance of approximately 0.25 mm, shown by the shaded region in Fig. 3d. This indicates that the homogeneous reaction rate and devolatilization rate maintain roughly a balance at a stand-off distance of approximately $2d_P$ for single particle diffusion flames in this study.

2.5. Particle number density evaluation

The pre-processing of DBI images included background subtraction and binarization using an adaptive threshold method [17] and the circle-equivalent diameter was calculated with respect to the particle area. The particle overlap was mostly avoided by limiting the particle seeding rate. For the purpose of evaluating the local particle number density (PND), a reference volume was defined as follows. First, a mean image was computed using the pre-processed projected particle images which was subsequently smoothed and binarized. The

binarized area guaranteed to include 95% of all particles. Then, the boundaries of the resulting area were approximated using a third-order polynomial fit which is superposed with an instantaneous binary image in Fig. 4a. The centerline had a maximum slant angle of 3.1° to the y-axis due to a slight misalignment error of the jet tube. The 2D reference area was projected to the x-y plane with respect to an angle of 17° between OH-LIF and DBI cameras. A 3D reference volume of the jet-like structure V_{jet} in Fig. 4b was computed by assuming rotational symmetry. Finally, PND is calculated as $PND = N/V_{jet}$ where N specifies the amount of particles.

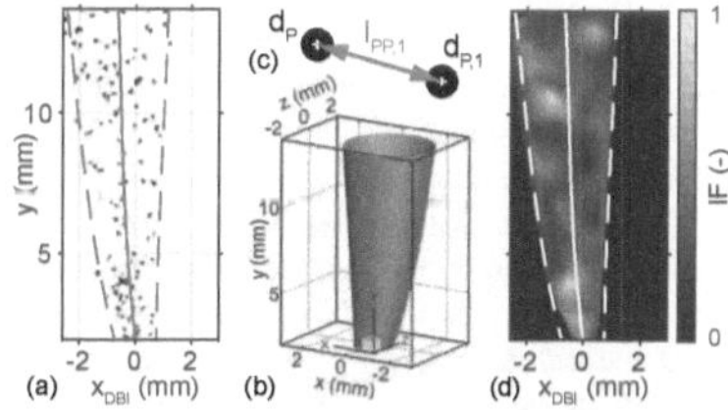

Figure 4: (a) Binarized instant DBI image. The boundaries and centerline of the particle jet are depicted by red lines. (b) Three-dimensional visualization of the particle jet. (c) IF calculation scheme for IF = 0.25. (d) Corresponding 2D map of IF.

Additionally, the particle-particle interaction was studied by introducing the 2D-projected interaction factor (IF) using $IF = 0.5 (d_P + d_{P,1}) / l_{PP,1}$. Here, $d_{P,1}$ denotes the individual circle-equivalent diameter of the nearest particle and $l_{PP,1}$ the instantaneous interparticle distance. Figure 4c illustrates the calculation schematic. IF = 1 indicates two touching particles and IF decreases with increasing $l_{PP,1}$ or decreasing particle diameters. Each particle is associated with an individual IF which is located at the centroid of its projected area. From this, a 2D distribution of IF is calculated using a triangulation-based natural neighbor interpolation shown in Fig. 4d.

3. Results and discussion

3.1. Flame topology transition

Figure 5 exemplifies four instantaneous group particle volatile flames represented by iso-contours of the OH-LIF intensity. Cross-sections through the flame's central plane are shown. The global PND is calculated within the entire 3D jet volume up to y = 13.7 mm (in gray) and the corresponding values are 0.18, 0.48,

4

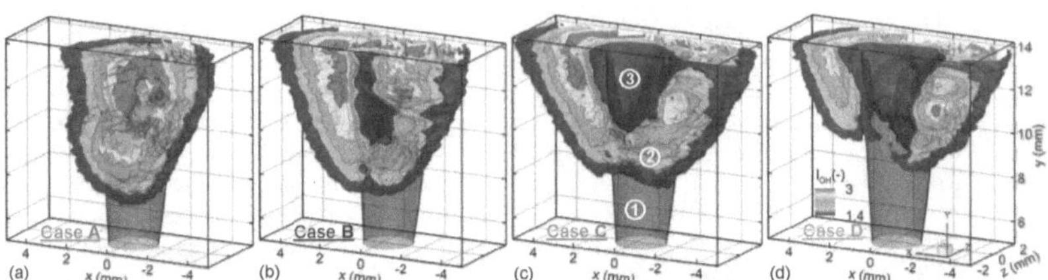

Figure 5: Flame topology transition of group particle combustion. From (a) to (d): instantaneous OH-LIF reconstruction of individual realizations A, B, C and D with global PNDs of 0.18, 0.48, 1.25 and 1.9 mm^{-3}, respectively.

1.25 and 1.9 mm^{-3} for the individual realizations A, B, C and D, respectively. In A, although an enclosed volatile flame is observed, spherical flame structures are still present in the vicinity of individual particles, which resemble single particle flames (see Fig. 3). Since the mean inter-particle distance l_{PP} is estimated > 1 mm, volatile combustion proceeds for individual particles and the burnt gas interlinks in the outer periphery. As the PND increases from B to D, a non-flammable region enclosed by the enveloping flame is observed which grows extensively with increasing particle number density. Simultaneous DBI images (not shown) confirm soot formation of particles located within this region. The flame structure extends wider and the ignition height increases. Near-spherical flames of individual particles, if present, appear only near the flame base and vanish at the high PND of D due to a complete inner non-flammable region. To systemize these observations, in the volume of interest three zones are classified: ① particle heating, ② volatile flame and ③ non-flammable region, which are marked in Fig. 5c.

To study the flame topology transition from dilute to dense particle groups, horizontal profiles of OH-LIF intensities in the central plane are analyzed for 4 and 5 mm downstream the LRF surface, in Fig. 6a and b, respectively. The profiles are normalized to the background intensity without particles and the laser absorption along the x-axis is corrected. The jet centerline is located at $x = 0$ mm. At $y = 4$ mm, the OH-LIF intensity near the centerline declines slightly as the PND increases from A to D. This distinction becomes markedly larger at 5 mm: while the intensity of A and B remains stable, C and D show a more distinct intensity drop. Considering the maximum particle heating rate of approximately 10^5 K/s and residence time of approximately 5 ms, the particle temperature is estimated < 800 K. Since the particle temperature decreases with an increasing PND, it remains below the ash melting temperature. No significant volatile concentration is expected due to the relative low temperature of particles [14]. Thus, the OH-LIF intensity is dominated by the local gas temperature. The transition process can be better explained by involving a schematic to illustrate realizations C (left) and D (right) in Fig. 6c.

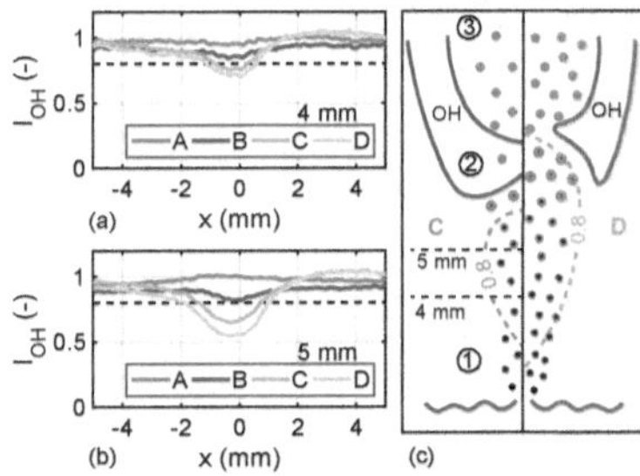

Figure 6: (a) and (b) Horizontal OH-LIF intensity profiles of background exhaust gas at 4 and 5 mm. (c) A schematic visualization of the particle group flame transition from C to D.

For the individual realization C, when the particles enter the LFR with high ambient temperature, heat transfer occurs from the gas phase to the particles resulting in a rise of the particle temperatures and a decrease of the gas temperature. The heat loss of the gas phase accumulates over time, as deduced from Fig. 6a and b. Thus, a region of reduced temperature forms around the particle group, which is indicated by the 0.8 iso-line of OH-LIF intensity (dashed line). Within this area, temperatures below 1400 K are estimated based on the OH-LIF intensity decrease considering the ambient temperature of 1800 K [4]. Particles near the jet boundary are substantially heated and are expected to release volatiles faster. A flammable mixture with the proper volatile mole fraction and high temperature oc-

5

curs therefore at first near the jet boundary. Shortly after that, auto-ignition occurs and forms an enclosing flame base. As the devolatilization of particle groups proceeds, the gas mixture becomes rich in volatiles and exceeds the flammability limits in zone ③. This effect is partially accelerated by the high temperatures across the flame in zone ②. Substantial outgassing of volatile matter pushes the diffusion flame away from the jet centerline (Fig. 5c). Without turbulent mixing, the diffusion flame is restricted by the diffusivity of oxygen and volatile matter with respect to each other. For realization D with higher PND, the low temperature region extends and penetrates the flame. The OH-LIF intensity profiles indicate an even lower temperature than C in zone ②. Numerical investigations have shown that for increasing PND, the gas temperature decreases which leads to lower particle heating rates and therefore lower devolatilization rates [13]. Even assuming that a region of mixture fraction within the flammability limits is available, the flame is still quenched near the particle jet centerline due to extensive heat losses. This causes the flame to be completely isolated from the low temperature region and the flame warps around the particle group at the periphery (Fig. 5d).

3.2. Particle velocity

The particle velocity is evaluated using a PIV-PTV combined algorithm (Davis 10, LaVision). The PIV algorithm provides a preliminary estimation of the particle velocities using binary DBI particle image sequences. However, the interrogation window size of 32×32 pixels intends to contain at least one single particle. For the subsequent PTV evaluation, a correlation window size of 16×16 pixels is applied with a 2 pixel tolerance relative to the PIV reference.

The statistical velocity evaluation is conditioned on the global PND of instantaneous DBI images. Figure 7a shows the particle velocity magnitude V_n over the y-axis for all instants with a similar PND to A to D. The axial component is dominant in V_n and the radial component is negligible under the laminar condition in this study. The initial V_n decreases with increasing PND because 1) partly inelastic particle collisions reduce the kinetic energy of the particle group in the injection tube, and 2) thermal expansion weakens due to low gas temperature in the post-flame region. The maximum ambient gas velocity is 1.5 m/s (dashed line). The V_n increases along the y-axis downstream since a slip velocity accelerates particles. Figure 7b shows V_n of realization C along the x-axis and at y-locations of 4, 5, 8 and 11 mm. The x-coordinate is referenced to the particle jet centerline (Fig. 4) and normalized to the particle jet radii

r_{jet}. The particles near the jet boundary exhibit a higher velocity than those near the centerline. Particles are accelerated by the higher velocity of the surrounding gas when approaching the high-temperature volatile flame.

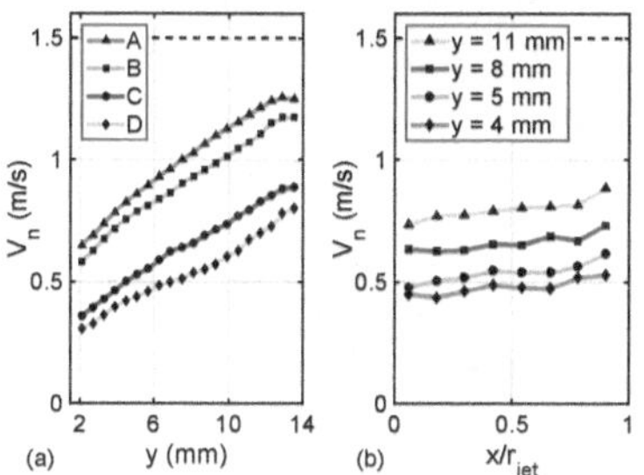

Figure 7: (a) Particle velocity magnitude V_n along the y-axis for PNDs comparable to A to D. (b) V_n along the x-axis and at several y-locations for C. The x-coordinate is referenced to the particle jet centerline and normalized by the particle jet radii r_{jet}.

3.3. Ignition height and ignition delay time

The evaluation of the ignition delay time includes several processing steps: 1) binarization of 3D flame structures using a fixed threshold, 2) extraction of the ignition height H_{ign} from the lowest location of the binary flame structure, 3) calculation of the mean particle velocity at different y-slices (bin width = 0.6 mm), and 4) accumulation of the particle flight time through each y-slice until H_{ign} to obtain the ignition delay time t_{ign}. The reference time t_0 is defined when particles pass the flame front location at y = 1.8 mm. The intensity threshold of 1.4 intends to deal with the intensity fluctuations of the premixed flat flame. H_{ign} and t_{ign} are statistically analyzed over PND_{ign}, which is a local PND restricted to the region of $H_{ign} \pm 0.5$ mm along the y-axis. Here, 1720 instantaneous flame structures are used in total and PND_{ign} spans over a range of $0 - 2$ mm^{-3}.

Figure 8a shows the scatter plot between H_{ign} over PND_{ign}. The mean H_{ign} curve is evaluated by slicing PND_{ign} with a bin width of 0.12 mm^{-3}. The gray shaded area and error bars highlight the region within ± 1 standard deviation. For dilute particle streams, the interparticle distances are large and H_{ign} is dominated by the ignition of individual particles. The apparent variation of the ignition location is a consequence of the scattered particle size and velocity distribution which affect the heating rate. With a further increase of PND_{ign}, particle groups show elevated ignition locations. H_{ign} overall increases from for individual realizations A to D, which

6

is in accordance with the observation of the flame topology transition in Fig. 5.

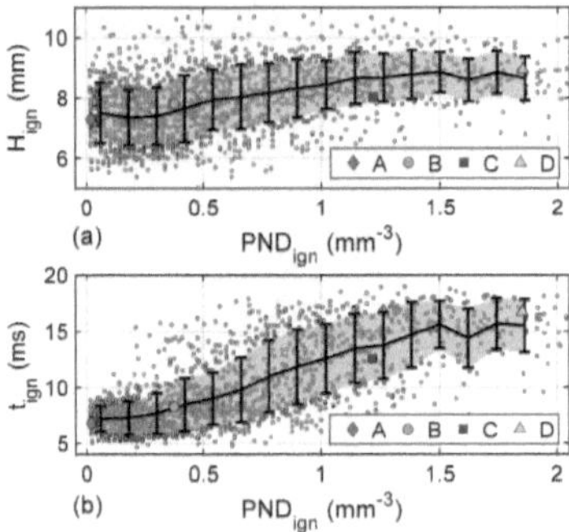

Figure 8: (a) Ignition height H_{ign} over local PND_{ign} associated with ignition. (b) Ignition delay time t_{ign} over PND_{ign}. Individual realizations A, B, C and D are highlighted.

Figure 8b depicts the relation between t_{ign} and PND_{ign}. Taking the particle velocity into account, for dilute particle densities, a small variation of t_{ign} is observed, which is mostly concentrated in the region of 6 - 9 ms. This result is in accordance with previous measurements of single particle ignition [4], hence confirms the dominance of single particle combustion characteristics. Due to the combined effects of gas temperature, particle heating rate and gas diffusivity, as aforementioned in section 3.1, t_{ign} increases monotonically towards higher particle number densities.

3.4. Non-flammability

The appearance of non-flammable regions in the flame interior allows studying the transition from single to group particle combustion. Referring to Fig. 5, the non-flammability volume V_3 of zone ③ and the effective flame volume V_2 of zone ② are evaluated using binary 3D flame structures. For flames with unconnected structures (i.g. individual realization D), a convex hull enclosing the flame within the measurement volume is defined. The non-flammability ratio R_{nf} is defined as $R_{nf} = V_3 / (V_3 + V_2)$. The analysis of R_{nf} is conditioned on PND_{flame}, which denotes the local particle number density in the region from the instantaneous ignition height to 13.7 mm above the LFR surface. The latter is the upper limit of PND_{flame} data restricted to the FOV of the DBI measurement. For all 1720 instantaneous flame structures investigated, 40% exhibit a non-flammable volume and are shown in Fig. 9a. The remaining 60% show full flammability with $R_{nf} = 0$ and

have a mean PND_{flame} of $0.37\,\text{mm}^{-3}$ (dashed line). The solid line shows the mean R_{nf} curve and individual realizations A to D are highlighted.

The OH-LIF signals are not measurable if the laser is absorbed by soot or blocked by particles within the laser sheets. This results in artificial R_{nf} despite low particle number densities. The maximum of artificial R_{nf} is estimated up to 0.003. Figure 9a shows that R_{nf} values are small and have a narrow distribution between 0 and 0.1 when $PND_{flame} < 0.37\,\text{mm}^{-3}$. This border represents the minimum particle number density of observed non-flammable regions. The growth of the non-flammability region is evident as PND_{flame} increases to $1.5\,\text{mm}^{-3}$.

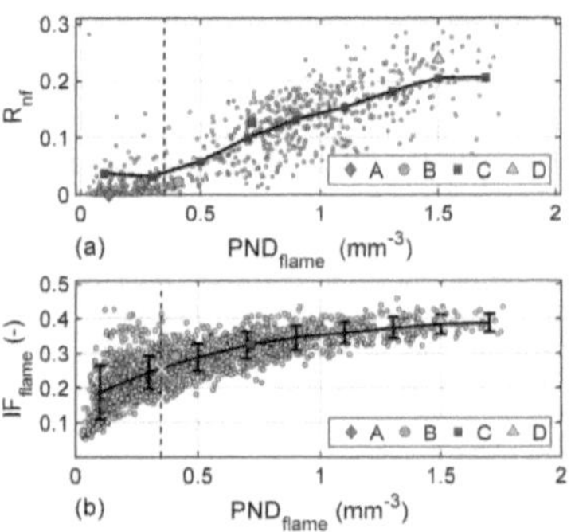

Figure 9: (a) Non-flammability volume ratio R_{nf} over PND_{flame}. (b) Correlation between PND_{flame} and IF_{flame}. The dashed line marks the mean PND_{flame} of flame structures with $R_{nf} = 0$. Individual realizations A, B, C and D are highlighted.

Potential interactions between particles are characterized by IF_{flame}, which is analyzed in the same region as PND_{flame}. Figure 9b presents the positive correlation of IF_{flame} and PND_{flame} for all instants. The mean curve of IF_{flame} and error bars of ± 1 standard deviation are shown (solid line). The minimum particle density for non-flammable regions (vertical dashed line) with its projected point (yellow cross) on the mean curve is marked. This value corresponds to an interaction factor $IF_{flame} = 0.25$, which indicates that the inter-particle distance is $4d_P$ or approximately 0.5 mm for the present mean particle diameter. Referring to single particle combustion, the reaction zone of a single particle can only sustain if the inter-particle distance l_{PP} is larger than twice the stand-off distance, namely 0.5 mm in this investigation. The non-flammable region in particle groups downstream was discussed in [13, 14] in the context of mixture fraction and gas temperature. In

7

the present study, however, transition to group combustion is deduced from interaction factors and particle number densities. Exceeding a specific limit (here $IF_{flame} > 0.25$ or $PND_{flame} > 0.37\,mm^{-3}$), the occurrence of non-flammability volume becomes more pronounced. Although different boundary conditions were used in [14], similar effects of lower gas temperatures and the appearance of non-flammable regions were observed with decreasing inter-particle distance.

4. Conclusions

The transition effects from single to group combustion of hvb coal particles under a laminar condition are investigated using simultaneous volumetric OH-LIF imaging and diffuse backlight-illumination measurements. 3D reconstructed volatile flames revealing distinct structural characteristics, which are extensively affected by the decrease of local gas temperatures with increasing PND, are visualized. Based on this, parametric variations of PND covering the range from individual particle combustion to clear group combustion are performed. The inter-particle and particle-gas interaction result in lower particle velocities V_n compared to single particles under the same conditions. Considering varying flight time with increasing PND, particle groups reveal a monotonic increase (up to a PND of $1.5\,mm^{-3}$) of ignition delay time. The appearance of non-flammable regions in the flame interior is quantified by evaluating the volume ratio R_{nf} over the local PND in the flame-relevant region. The non-flammability becomes pronounced if PND exceeds a limit of approximately $0.37\,mm^{-3}$. Correlations between PND and IF suggest that the critical inter-particle distance corresponds to about $4d_P$ in this study. This criterion might change by using other types or differently sieved particles, while analogous effects are expected regarding the transition from single to group combustion. Current results provide a novel data set for numerical studies of coal particle group combustion under oxy-fuel relevant conditions.

References

[1] K. Annamalai, W. Ryan, S. Dhanapalan, Interactive processes in gasification and combustion—Part III: Coal/char particle arrays, streams and clouds, *Prog. Energy Combust. Sci* 20 (1994) 487–618.

[2] Y. Yuan, S. Li, G. Li, N. Wu, Q. Yao, The transition of heterogeneous–homogeneous ignitions of dispersed coal particle streams, *Combust. Flame* 161 (2014) 2458–2468.

[3] C. R. Shaddix, A. Molina, Particle imaging of ignition and devolatilization of pulverized coal during oxy-fuel combustion, *Proc. Combust. Inst.* 32 (2009) 2091–2098.

[4] J. Köser, T. Li, N. Vorobiev, A. Dreizler, M. Schiemann, B. Böhm, Multi-parameter diagnostics for high-resolution in-situ measurements of single coal particle combustion, *Proc. Combust. Inst.* 37 (2019) 2893–2900.

[5] J. Köser, L. G. Becker, A.-K. Goßmann, B. Böhm, A. Dreizler, Investigation of ignition and volatile combustion of single coal particles within oxygen-enriched atmospheres using high-speed oh-plif, *Proc. Combust. Inst.* 36 (2017) 2103–2111.

[6] L. Zhang, E. Binner, Y. Qiao, C.-Z. Li, High-speed camera observation of coal combustion in air and O_2/CO_2 mixtures and measurement of burning coal particle velocity, *Energy Fuels* 24 (2010) 29–37.

[7] L. Zhang, E. Binner, Y. Qiao, C.-Z. Li, In situ diagnostics of victorian brown coal combustion in O_2/N_2 and O_2/CO_2 mixtures in drop-tube furnace, *Fuel* 89 (2010) 2703–2712.

[8] A. C. Sarroza, T. D. Bennet, C. Eastwick, H. Liu, Characterising pulverised fuel ignition in a visual drop tube furnace by use of a high-speed imaging technique, *Fuel Processing Technology* 157 (2017) 1–11.

[9] W. Weng, M. Costa, Z. Li, M. Aldén, Temporally and spectrally resolved images of single burning pulverized wheat straw particles, *Fuel* 224 (2018) 434–441.

[10] B. Goshayeshi, J. C. Sutherland, A comparison of various models in predicting ignition delay in single-particle coal combustion, *Combust. Flame* 161 (2014) 1900–1910.

[11] R. Knappstein, G. Kuenne, A. Ketelheun, J. Köser, L. Becker, S. Heuer, M. Schiemann, V. Scherer, A. Dreizler, A. Sadiki, J. Janicka, Devolatilization and volatiles reaction of individual coal particles in the context of fgm tabulated chemistry, *Combust. Flame* 169 (2016) 72–84.

[12] G. L. Tufano, O. T. Stein, A. Kronenburg, G. Gentile, A. Stagni, A. Frassoldati, T. Faravelli, A. M. Kempf, M. Vascellari, C. Hasse, Fully-resolved simulations of coal particle combustion using a detailed multi-step approach for heterogeneous kinetics, *Fuel* 240 (2019) 75–83.

[13] S. Farazi, J. Hinrichs, M. Davidovic, T. Falkenstein, M. Bode, S. Kang, A. Attili, H. Pitsch, Numerical investigation of coal particle stream ignition in oxy-atmosphere, *Fuel* 241 (2019) 477–487.

[14] G. L. Tufano, O. T. Stein, B. Wang, A. Kronenburg, M. Rieth, A. M. Kempf, Coal particle volatile combustion and flame interaction. Part I: Characterization of transient and group effects, *Fuel* 229 (2018) 262–269.

[15] T. Li, J. Pareja, L. Becker, W. Heddrich, A. Dreizler, B. Böhm, Quasi-4d laser diagnostics using an acousto-optic deflector scanning system, *Appl. Phys. B* 123 (2017) 1243.

[16] T. Li, B. Zhou, J. H. Frank, A. Dreizler, B. Böhm, High-speed volumetric imaging of formaldehyde in a lifted turbulent jet flame using an acousto-optic deflector, *Experiments in Fluids* 61 (2020) 2903.

[17] J. Pareja, A. Johchi, T. Li, A. Dreizler, B. Böhm, A study of the spatial and temporal evolution of auto-ignition kernels using

8